SIMULATION AND ITS DISCONTENTS

SIMPLICITY: DESIGN, TECHNOLOGY, BUSINESS, LIFE

John Maeda, Editor

The Laws of Simplicity, John Maeda, 2006

The Plenitude: Creativity, Innovation, and Making Stuff, Rich Gold, 2007

Simulation and Its Discontents, Sherry Turkle, 2009

SIMULATION AND ITS DISCONTENTS

SHERRY TURKLE

With additional essays by
William J. Clancey, Stefan Helmreich, Yanni A. Loukissas,
and Natasha Myers

The MIT Press
Cambridge, Massachusetts
London, England

For information about special quantity discounts, please email special_sales@ mitpress.mit.edu.

This book was set in Scala and Scala Sans by Graphic Composition, Inc., Bogart, Georgia. Printed and bound in the United States of America.

Library of Congress Cataloging-in-Publication Data

Simulation and its discontents / Sherry Turkle ; with additional essays by William J. Clancey . . . [et al.]. ; foreword by John Maeda

   p. cm. — (Simplicity—design, technology, business, life)

Includes bibliographical references and index.

ISBN 978-0-262-01270-6 (hbk. : alk. paper)

1. Computer simulation. 2. Visualization. 3. Technology—History. 4. Technology—Social aspects. I. Turkle, Sherry.

QA76.9.C65T87   2009   003′.3—dc22   2008035982

10 9 8 7 6 5 4 3 2 1

To David Riesman and Donald Schön,
Two mentors, two friends

CONTENTS

CONTENTS

FOREWORD

John Maeda

In the mid-1800s it was rumored that there was gold in "them thar hills" of California, and hundreds of thousands of settlers flocked to stake out their fortunes. There's nothing like a plum, virgin opportunity to attract the curious. The world sometimes presents us with new possibilities, and either we jump at the chance . . . or just stay at home.

The World Wide Web was another kind of gold rush when it first got started. I was one of those people that shrugged unconvincingly when excited friends with a kind of "gold fever" written across their faces would exhort to me, "You have to have a homepage!" "A homepage?" I would mockingly question, "I already have a home. No, thank you." So I stayed at home at first. But jumped in soon thereafter. Luckily.

Today I spend most of my time working diligently on the Web, and at least once a week I buy some "whatever.com" domain name as is the preferred personal hobby of the armchair digital magnate. So having jumped on the Web bandwagon just in time to not lose out, I vowed to not be so aloof when the next trend would come around. And then the opportunity presented itself—called "Second Life."

I bring up Second Life because it embodies Turkle's thesis that we are amidst a long-running sea change in the world of "virtual worlds" that has not yet become manifest as a switchover. Faithful to the gold rush mentality and remembering my doubts in the Web, at the peak of the buzz I bought an island in Second Life for a few thousand dollars. Five months later I gave it to a friend as I had no use for it. Reading the various strands of historical discontent as documented by Turkle, I now know why I let my island go. And I also know that some day I will wish I had it back.

For businesses that wish to understand the importance of understanding the history of doubt that surrounds adoption of simulated realities, just look at the current popularity of Facebook to viscerally feel that something is indeed happening in our world that looks like a mass migration to virtuality. The concepts in this short historical piece by Turkle trace a clear path from the technology in the laboratory to the living virtual landscape in which we thrive today. "Knowledge makes things simpler," by the fourth law of simplicity, and for that reason I know the knowledge in this volume will certainly untangle some of the mysteries you felt, as I once did, about the design and technology issues that pertain to the simulations in which we will all live.

For Susan Sontag, "to collect photographs is to collect the world."[1] She spoke about the "grandiose" result of the photographic enterprise: it provides the sense that we "hold the whole world in our heads."[2] Simulation takes us further into our representations. We no longer need to keep the world in our "mind's eye." We build it, step into it, manipulate it. If photography is a new way of seeing, simulation is that and more: a new way of living, both a change of lens and a change of location.

"Simulation and Its Discontents" draws on two ethnographic studies for which I was principal investigator. The first, sponsored by the MIT provost's office, explored the introduction of intensive computing into educational practice at MIT in the mid-1980s; the second, a National Science Foundation study twenty years later, investigated simulation and visualization in contemporary science, engineering, and design.[3] "Simulation and Its Discontents" is followed by four case studies, under the rubric "Sites of Simulation." Two of the cases, those by Yanni A. Loukissas and Natasha Myers, grew out of the NSF study of simulation and visualization; the cases by Stefan Helmreich and William J. Clancey were commissioned for this book. Together, we give voice to how scientists, engineers, and designers have responded to simulation and visualization technologies as these became central to their work over the past twenty-five years.[4]

I gratefully acknowledge my colleagues who collaborated on the studies of the mid-1980s and mid-2000s. My coprincipal investigator on the study of educational computing at MIT was Donald Schön. From 1983 to 1987 we worked with research assistants Brenda Nielsen (who did fieldwork in the Department of Architecture, Chemistry, and Physics), M. Stella Orsini (who did fieldwork in the Department of Architecture and Chemistry), and Wim Overmeer (who did fieldwork in the Department of Civil Engineering). On the second study, from 2002-2005, I worked with coprincipal investigators Joseph Dumit, Hugh Gusterson, David Mindell, and Susan Silbey. This study was part of the research effort of the MIT Initiative on Technology and Self. Gusterson contributed materials on nuclear weapons design, Mindell on the history of aviation. Mindell supervised Arne Hessenbruch, a research assistant who helped establish an overview of our project. Dumit and Silbey supervised Natasha Myers, a research assistant, in her case study of the life sciences; I worked with research assistant Yanni A. Loukissas on a case study of architecture. The NSF research was integral to Loukissas's and Myers's dissertation projects. I am particularly in debt to these two talented scholars who were involved in every aspect of researching and writing the NSF final report.

In addition to field research, the NSF project supported two workshops on simulation and visualization in the professions, one in Fall 2003 and one in Spring 2005. Each brought together scientists, engineers, and designers from a range of disciplines. I thank the informants from the 1980s and the 2000s, those who allowed us to watch them at work, those who shared their thoughts in individual interviews, and those who participated in workshop discussions. In "Simulation and Its Discontents" all have been given anonymity. Where a name appears, it is a pseudonym. All of the case study material follows this policy as well.

My work on simulation and contemporary professional life has also been supported by the MIT Program in Science, Technology, and Society; the Mitchell Kapor Foundation; the Intel Corporation; and the Kurzweil Foundation through their support of the MIT Initiative on Technology and Self. This volume owes a debt to the collegial life of the Initiative, to the MIT Program in Science, Technology, and Society, and to the MIT Media Laboratory. I gratefully acknowledge the contributions of Anita Say Chan, Jennifer Ferng, William D. Friedberg, William J. Mitchell, William Porter, Rachel Prentice, and Susan Yee. Ferng worked as a research assistant on developing the history of simulation in architectural practice; Prentice documented the first MIT workshop of simulation and visualization in Fall 2003; Friedberg provided an insightful critical reading; Chan, Mitchell, Porter, and Yee helped me to think through the tension between doing and doubting that became central to my thinking.

Kelly Gray brought this volume her tenacity and talent; she is that ideal reader—interested, knowledgeable, and critical—that every author hopes to find. This is the fourth in a series of MIT Press books that grew out of work at the Initiative on Technology and Self. Gray was committed to the Initiative and this ambitious publication project from the very start; both are in her debt.

I thank Judith Spitzer and Grace Costa for providing the administrative support that enabled me to do my best work. And I am grateful to the wonderful people at the MIT Press who have worked with me through the Initiative's publication project—Deborah Cantor-Adams, Erin Hasley, and Robert Prior. With this book, Margy Avery, Erin Shoudy, and Sharon Deacon Warne joined this group of colleagues who always make things better.

When I first dreamed of an Initiative on Technology and Self at MIT, my daughter Rebecca was eight. As I write these words, she is applying to college. Having her in my life as the Initiative flourished has been a great gift. I often tell the story of Rebecca, at eight, sailing with me on a postcard-blue Mediterranean, shouting "Look, Mommy, a jellyfish! It's so realistic!" as she compared what she saw in the water to the simulation of sea creatures she had so often seen on her computer at home. For Rebecca and her friends, simulation is second nature. I wanted to write a book that would make it seem rather less so, reminding them that it brings new ways to see and to forget.

Sherry Turkle
Provincetown
Summer 2008

SIMULATION AND ITS DISCONTENTS

Sherry Turkle

WHAT DOES SIMULATION WANT?

It was September 1977, my first week on the faculty at MIT. Trained as a psychologist and sociologist, I was finding my bearings in a sea of scientists, engineers, and designers. A colleague in civil engineering took me to lunch to give me the lay of the land. He jokingly told me that I had come at a good time but had missed a golden age: "This place is going to hell."[1] At the heart of the decline as he saw it: students used calculators instead of slide rules. With slide rules, he explained, the user had to know the number of decimal places that made for a meaningful answer. With calculators, this was no longer required. Students, he reported, had lost all sense of scale. In his classes, answers were coming back wrong by orders of magnitude. Moreover, students couldn't manipulate numbers in their heads the way they used to. "And the calculator thing is small potatoes," he said. We spoke of the new personal computers, only recently on the scene; he saw them as giant calculators. Projecting forward, he unhappily imagined computers in pedagogy. Scientists and engineers had to have "numbers in their fingers . . . the back of the envelope calculation is where science is born." He told me to keep my eyes open for the kinds of change that come once in a lifetime.

Six years later, I was studying faculty and student reactions to the widespread introduction of personal computers to the undergraduate MIT experience (an initiative known as "Project Athena"); computers were now officially central to pedagogy. Twenty years after that, I was investigating how simulation and visualization had changed the face of research and teaching in science, engineering, and design. At a Fall 2003 MIT workshop on this theme, an MIT molecular biologist offered an arresting gloss on the fears expressed at lunch a quarter century before. He admitted that his students "couldn't tell the difference between x to the 12th power or x to the 24th." He went on to say that naturally this couldn't be a good thing, but "what they do know how to do, is hit the calculator button twice to make sure they get the right answer."

This biologist builds mathematical representations of molecules for virtual experiments. His goal is to build a scientific life in simulation. He is a major contributor; his laboratory is changing our understanding of proteins. Neither MIT nor protein science has gone to hell. But the civil engineer who took me to lunch in Fall 1977 had a point: I had been witness to a sea change.

"Simulation and Its Discontents" is my view of that sea change. My studies of the 1980s and 2000s explored simulation as a dominant force in changing scientific and design identities.[2] Here I trace the threads of doubt raised by those I met along the way. Why focus on discontents? These days we see the world through the prism of simulation. Discontents with this hegemony draw our attention to settings where simulation demands unhappy compliance; discontents draw our attention to things that simulation leaves out. As is the case when we study scientific controversy, looking at discontents is a way to discover deep commitments.[3]

Among my subjects of the 1980s were custodians of doubt. They were professors at MIT who, for the most part, saw simulation as central to the future of their disciplines. Yet, as they introduced it to their students, they were sensitive to the ways it could overreach. In prospect, they saw creativity, certainly, but also the opportunity for seduction and betrayal, times when simulation might beguile. They feared that even skeptical scientists would be vulnerable to the allure of a beautiful picture, that students would be drawn from the grittiness of the real to the smoothness of the virtual. These days, professionals who voice discontent about simulation in science, engineering, and design run the risk of being seen as nostalgic or committed to futile protest. The early skeptics may have felt they were engaged in what one called a "rear-guard action," but they did not feel their objections to be futile. They worked in the attractive belief that they could take action to protect what was important. They wanted to preserve what they termed "sacred spaces," places where technology might disrupt sacrosanct traditions linked to core values. So, for example, architects wanted to preserve hand drawing; they stressed its history, its intimacy, and how it tied architecture to the arts. Physicists wanted to maintain the pedagogy of the lecture hall because they saw it as a place to model a scientific identity. A physicist in a lecture hall was there to answer such questions as: What do physicists care about? What do they put aside? How do they handle doubt?

Twenty years later, in the study of simulation and visualization in the 2000s, I hear echoes of these early discontents, now voiced by some of simulation's most sophisticated practitioners. Across disciplines, there is anxiety about the retirement of senior colleagues: they are seen as special because they were in touch with a way of doing science and design that was less mediated, more direct. The senior colleagues used pencils; they knew how to revise drawings by hand; in the laboratory, they knew how to build and repair their own instruments. They understood computer code, and when things weren't working right, they could dive into a program and fix it from the ground up. As they retire, they take something with them that simulation cannot teach, cannot replace.

Sensibilities shift. In the 1980s, an MIT engineering student was amazed to learn that there was a time when skyscrapers were designed without computers. He could not imagine how engineers could tackle such projects "by hand." To this student, a 1950s highrise was a veritable pyramid, even prehistoric, life before simulation. Twenty years ago, professionals in science and design flirted with simulation even as they were suspicious of it. Today, they are wary but wed to it.

In a design seminar, the master architect Louis I. Kahn once famously asked: "What does a brick want?"[4] It was the right question to open a discussion on the built environment. Here, I borrow the spirit of this question to ask, "What does simulation want?" On one level, the answer to this second question is simple: simulations want, even demand, immersion. Immersion has proved its benefits. Architects create buildings that would not have been imagined before they were designed on screens; scientists determine the structure of molecules by manipulating them in virtual space; nuclear explosions are simulated in 3D immersive realities; physicians practice anatomy on digitized humans.[5]

Immersed in simulation, we feel exhilarated by possibility. We speak of Bilbao, of emerging cancer therapies, of the simulations that may help us address global climate change. But immersed in simulation, we are also vulnerable. Sometimes it can be hard to remember all that lies beyond it, or even acknowledge that everything is not captured in it. An older generation fears that young scientists, engineers, and designers are "drunk with code." A younger generation scrambles to capture their mentors' tacit knowledge of buildings, bodies, and bombs. From both sides of a generational divide, there is anxiety that in simulation, something important slips away.

In 1984 an MIT professor of architecture said that to use simulation responsibly, practitioners must learn "to do" and "to doubt." He thought that students were not in a position to sufficiently doubt simulation, because the demands of acquiring technical mastery made it too hard to achieve critical distance. But he believed that, in the end, professional maturity would bring with it both immersion and skepticism.

Things have not been so simple. Simulation makes itself easy to love and difficult to doubt. It translates the concrete materials of science, engineering, and design into compelling virtual objects that engage the body as well as the mind. The molecular model built with balls and sticks gives way to an animated world that can be manipulated at a touch, rotated, and flipped; the architect's cardboard model becomes a photorealistic virtual reality that you can "fly through." Over time, it has become clear that this "remediation," the move from physical to virtual manipulation, opens new possibilities for research, learning, and design creativity.[6] It has also become clear that it can tempt its users into a lack of fealty to the real.[7] With these developments in mind, I returned to the experiences of simulation's early adopters with new regard for their anxieties as well as their aspirations. At the heart of my story is the enduring tension between doing and doubting. Simulation demands immersion and immersion makes it hard to doubt simulation. The more powerful our tools become, the harder it is to imagine the world without them.

THE VIEW FROM THE 1980s

Winston Churchill once said, "We make our buildings and afterwards they make us. They regulate the course of our lives."[1] We make our technologies and our technologies shape us. They change the way we think. Recently, during a student conference, an MIT professor of architecture spoke of designing his first building, a small lakeside cottage, when he was fresh out of graduate school in the early 1960s: "The connection between drawing and thinking, of sketching until the building was in your blood—that is what made a great design." Later that week, in a workshop on simulation, a senior architect remarked on being taken aback when he realized that his current design students did not know how to draw. They had never learned; for them, drawing was something the computer could do. His response was to insist that all his students take drawing classes. But he had no illusions. This new generation of students, all fluent users of simulation, is well on its way to leaving behind the power of thinking with a pencil.

From one point of view, the story of an architect who sketches with a computer program rather than with pencil and paper is a story about individual creative experience. It might be about loss or new possibilities. It might be about both. But professional life takes place in public as well as private spheres; the story of the architect and the pencil is about social consequences and aesthetics. Life in simulation has ethical and political as well as artistic dimensions.[2]

Individuals become immersed in the beauty and coherency of simulation; indeed simulations are built to capture us in exactly this way. A thirteen-year-old caught up in *SimCity*, a game which asks its users to play the role of urban developers, told me that among her "Top Ten Rules of *Sim*" was rule number 6: "Raising taxes leads to riots." And she thought that this was not only a rule of the game but a rule in life.[3] What may charm in this story becomes troubling when professionals lose themselves in life on the screen. Professional life requires that one live with the tension of using technology and remembering to distrust it.

PROJECT ATHENA: DESIGN IN THE GARDEN

Project Athena, launched in 1983, brought personal computing to an MIT education.[4] Across all disciplines, faculty were encouraged, equipped, and funded to write educational software to use in their classes. For many, the Athena experience marked a first brush with simulation and visualization technologies.

The experience could be heady. In the School of Architecture and Planning, some faculty saw computing as a window onto new ways of seeing. To architects accustomed to thinking with pencils in hand, designing on screens suggested novel ways of envisioning space and thinking about the process of design. Students could, as one design professor put it, "go inside programs" to "construct, change, interpret, and understand" how their models worked. Members of the architecture faculty dreamed about students moving from an aerial to a "worm's-eye" view of their sites, of moving from two-dimensional representations into virtual buildings that could come to life.

In the 1980s, this kind of talk was visionary. At that time, faced with slow, clunky equipment, it was not obvious that computers would drive new design epistemologies. A generation had grown up using "batch processing"—a style of computing where you dropped off data at the computer center and picked it up a day later. It was a time when screens were most associated with television, and indeed the same professors who imagined computers changing the basic elements of design also worried that computers might lull designers into the passive habits that came from TV viewing. Interactivity was a word that was only starting to become associated with computers.

The Athena story brings us back to a moment when educators spoke as though they had a choice about using computers in the training of designers and scientists.[5] Most of their students knew better. While his professors were still typing journal articles on typewriters, an MIT architecture student admitted that once he was given access to word processing and a fancy laser printer, not using them made him feel "stupid . . . behind." Another surprised her professors by sharing her idea that computers would soon "be like phones" and part of an architect's daily life.

Students also understood that although the Athena project was cast as an "experiment" and commonly discussed that way on campus, it was never really an experiment. One architecture student said that when she asked her professor why no one was trying to compare how structures designed "manually" actually "stacked up" against structures designed on the screen, she just got a shrug. As she put it: "It was just assumed that the computer would win." From the time it was introduced, simulation was taken as the way of the future.

In the School of Architecture and Planning it was clear from the early 1980s that graduates needed fluency with computer-aided design to compete for jobs. Athena's resources would make it possible to build a computer resource laboratory to meet their needs. This laboratory was called the "Garden," a name that celebrated and was celebrated by its inclusion of ceiling-high ficus trees.

A style of mutual support flourished under the ficus. Those students known to be experts were sought out by those who needed help. Everyone took turns; courtesy was cultivated; students came to know each other's projects. The Garden, open twenty-four hours a day, was one of the most heavily used computer facilities on campus. Even the early computer-aided design programs of the mid-1980s made it possible to move rapidly through a series of design alternatives and tinker with form, shape, and volume. Software presented "the defaults," predrawn architectural elements to manipulate. Faculty were amazed at how students, using them, came up with plans that were novel even to experienced designers.

Design students spoke glowingly of how easy it was to do multiple versions of any one design. Multiple passes meant that mistakes generated less anxiety. Mistakes no longer meant going back to the drawing board. Now mistakes simply needed to be "debugged" in simulation. In debugging, errors are seen not as false but as fixable. This is a state of mind that makes it easy to learn from mistakes.[6] Multiple passes also brought a new feel for the complexity of design decisions. Civil engineers as well as architects said that simulation let them go beyond simple classroom exercises to problems that gave them a sense of "how an actual building behaves under many conditions." For some, computer-aided design made theory come alive. One student observed that when he studied engineering principles in class, "I knew them to be facts, but I didn't know them to be actual and true." But when he saw the principles at work in virtual structures, he experienced them in a more immediate way. He expressed a paradox that will become familiar: the virtual makes some things seem more real.

But there were troubles in Eden. Most faculty in Architecture and Planning did not like to teach the details of software and programming in their classrooms. While this was in line with an MIT tradition of expecting students to master the technology required for coursework on their own,[7] it also reflected faculty ambivalence about bringing computers into the design curriculum in the first place. Many took the position that computer-aided design was only acceptable if practitioners were steeped in its shortcomings. Faculty referred to this as having a "critical stance." One professor didn't believe that students were in a position to achieve this balance because learning how to use simulation demanded a suspension of disbelief: "They can't be doubting and doing at the same time." And indeed, most students were overwhelmed by how hard it was to use the new technology; getting the details right took their total concentration. In general, students felt that if the faculty wanted them to have a critical stance toward simulation, then the faculty should teach it to them.

What emerged in Architecture and Planning was a division of labor. The students took up the doing; the faculty took up the doubting. For students, computer literacy became synonymous with technical mastery; for faculty, it meant using simulation even as one appreciated its limitations.

"OWNING" DESIGN

From the beginning of Athena, architecture faculty noted that computer fluency had its price, for example, design programs seemed to draw some students into sloppy, unconsidered work—making changes for change's sake. For their part, students admitted feeling captivated by software, at times so much so that they felt "controlled" by it. For one: "It is easy to let the simulation manipulate you instead of the other way around." Another student felt that the computer's suite of "defaults" constrained her imagination. When she drew designs by hand, she had a set of favorite solutions that she was likely to "plug in." Yet using preset computer defaults seemed different from these. For one thing, she was not their author. For another, they were too easy to implement. She said: "To get to the default you just have to hit return. It makes me feel too secure. If you use the default, it is as though the computer is telling you, 'this is right.'" One student asked of a default setting: "Why would it be in the computer if it wasn't a good solution?" Defaults offered feelings of instant validation that could close down conversation.

A student accustomed to planning her designs with cardboard models felt hemmed in by the computer programs she was forced to use in class. The cardboard had provided an immediate, tactile feedback she enjoyed; designing on the screen felt rigid. In 1984, this student was asked to view the design tools of her "youth" as relics. She was allowed to bring cardboard, plywood, paste, and dowels into her studio classes, but she was discouraged from using them until her designs had been "finalized" in simulation.

Another student summed up his experience of designing in simulation by saying that it had its good points but demanded "constant vigilance": "If you just look at the fine, perfect, neat, and clean sketches that come out of the machine and don't keep changing things . . . going back and forth between hand drawings and the computer, you might forget that there are other ways of connecting to the space."

Ted Randall, a professor of urban planning, lamented, "You love things that are your own marks. In some primitive way, marks are marks. . . . I can lose this piece of paper in the street and if [a day later] I walk on the street and see it, I'll know that I drew it. With a drawing that I do on the computer . . . I might not even know that it's mine. . . . People do analyses of their plan [on the computer] but they only fall in love with the marks they make themselves."

In the 1980s, the final products of computer-aided design programs lacked the artistry and personality of hand-drawn work. Some faculty were so demoralized by the aesthetics of computer printouts that they encouraged students in artisanal "compensations." For example, faculty suggested that students enhance the appearance of their computer projects by using colored pencils to "soften" the printouts. Softening played an emotional as well as aesthetic role. By making things seem more attractive and handcrafted, softening helped young designers feel more connected to their work. But for Randall, this attempt to bridge drawing and printout only provided ornamentation. What was being lost was the deep connection between hand and design.

Randall was wary of simulation because it encouraged detachment:

Students can look at the screen and work at it for a while without learning the topography of a site, without really getting it in their head as clearly as they would if they knew it in other ways, through traditional drawing for example. . . . When you draw a site, when you put in the contour lines and the trees, it becomes ingrained in your mind. You come to know the site in a way that is not possible with the computer.

He went on to tell the story of a student who, working on the computer, drew a road on a slope that was too steep to support it. The student had made this mistake because he had left out a contour line in his onscreen drawing, a contour line that represented twenty-five feet on the site. When Randall asked the student to explain what had happened, the student replied, "Well, it's only one contour." Given the screen resolution of the technology in the Garden, it would have been impossible to distinguish the lines on the screen if the student had put in the missing contour line. The student's screen could not accommodate more contour lines and the student had deferred to the technology. Randall commented:

It was only one contour, but [in physical space] that contour was twenty-five feet. The computer had led to a distortion of the site in the student's mind. He couldn't put more contour lines on the drawing, he said, because it was "too confusing." He said, "I couldn't work with that many contours. . . . I can't tell the lines apart any more."

In one reading of this vignette about an omitted contour line, it is a story about the limitations of a particular technology. The program was not able to "zoom in" to sufficient resolution; there was not "screen space" for the necessary number of contour lines. But Randall told the story to illustrate a larger concern. He believes that there will always be something in the physical real that will not be represented on the screen. Screen versions of reality will always leave something out, yet screen versions of reality may come to seem like reality itself. We accept them because they are compelling and present themselves as expressions of our most up-to-date tools. We accept them because we have them. They become practical because they are available. So, even when we have reason to doubt that screen realities are true, we are tempted to use them all the same.

Randall saw screens as shifting our attention from the true to what we might think of as the "true-here," the true in the simulation.

To Randall, his students' already noticeable allegiance to the "true-here," their lack of interest in questioning what was on the screen, was facilitated by the opacity of the tools used in computer-aided design. Despite early hopes that students would study programming and become fluent with the underlying structure of their digital tools, most were designing on systems whose inner workings they did not understand. If a tool is opaque to you, verification is not possible; in time, Randall worried, it may come to seem unnecessary. He was not alone in his concerns. One of his colleagues commented that in simulation, students "got answers without really knowing what's happened." Another felt that students using opaque tools would conceptualize buildings or cities "only in ways that can be embedded in a computer system." Students had no choice but to trust the simulations, which meant they had to trust the programmers who wrote the simulations.

In the early days of Athena, architecture and planning faculty tended to take sides for and against the computer. Critics talked about the danger of architecture becoming "mere engineering" and about the opacity of design programs. Proponents focused on new creative possibilities. In their view, simulation would enable design to be reborn.

Both critics and proponents felt the ground shifting; the fundamentals of design practice were being called into question. It was a thrilling, yet confusing time. And so, it is not surprising that many found themselves in conflict, defending hard-to-reconcile positions, for example, that computers would revolutionize design epistemology *and* that the heart of design would remain in the human domain. Faced with daily conflict, many designers found ways to dismiss the enormity of the change they expected. One strategy was to declare problematic effects inevitable but to relegate them to a far-distant future. Another was to minimize changes through rhetoric: for a while it seemed that the more a designer argued that the computer would bring revolutionary change, the more he or she invoked the idea of the computer as "just a tool." This phrase did a lot of work. Calling the computer "just a tool," even as one asserted that tools shape thought, was a way of saying that a big deal was no big deal.

In the early 1980s, even faculty who were most invested in the future of computer-aided design began to argue that an aspect of design education should be left untouched by the machine. This was drawing. Design software was being introduced that could serve as an alternative to drawing. Some faculty were pleased that now design professions would not exclude those without a gift for drawing. But others objected. They feared that if young designers relied on the computer, they might never achieve the intimate understanding of a design that comes from tracing it by hand. Designers described drawing as a sacred space, an aspect of design that should be kept inviolate. They pointed out that the practice of drawing, pencil on paper, made it clear that the touchstone for design was close to the body. They made it clear that even as new tools enable new ways of knowing, they also lead to new ways of forgetting.

DESIGN AND THE PROGRAMMER

In the 1980s, alternate visions of computers and the future of design were expressed in competing views about programming. Some architects believed that designers needed to learn advanced programming. If designers did not understand how their tools were constructed, they would not only be dependent on computer experts but less likely to challenge screen realities. Other architects disagreed. They argued that, in the future, creativity would not depend on understanding one's tools but on using them with finesse; the less one got tied up in the technical details of software, the freer one would be to focus exclusively on design. They saw programming and design as at odds; they discussed them as though the technicity of the first would impinge on the artistry of the second. Such views were influential: architecture students peppered their conversation with phrases such as, "I will never be a computer hacker" and "I don't consider myself a hacker." One graduate student in architecture explained how it was not possible to be both a "computer person" and a good designer at the same time: "Can you be a surgeon and a psychiatrist? I don't think you can. I think you have to make a choice." Another worried about becoming too competent at programming: "I don't want to become so good at it that I'm stuck in front of a computer forty hours a week. It's a matter of selective ignorance."

Civil engineering students were similarly divided on the question of programming. Some made it clear that if they had wanted to become programmers, they would have gone into computer science. They were happy that simulation tools, even if opaque, made it possible for them to attack complex problems from early on in their careers. But civil engineering students and faculty, like their colleagues in architecture, were also concerned about their growing dependency on the computer scientists who built their tools. A design program called Growltiger became a lightning rod for these concerns.

From 1986 on, Growltiger made it possible for civil engineers to create a visual representation of a structure, make a local modification to it, and explore the consequences of that move for the behavior of the structure as a whole. Students appreciated Growltiger because it released them from the burden of tedious calculation. They used it for theory testing, intuition building, and playful exploration. But, as when architects considered computer-aided design, enthusiasm for what this opaque program offered was tempered by anxieties about what it might take away. In particular, there was fear that ready-made software would blind engineers to crucial sources of error and uncertainty. Just as drawing became a sacred space for many MIT architects, something they wanted to maintain as a "simulation-free zone," so, too, civil engineers frequently talked about the analysis of structures as off-limits for the computer.

Civil engineering faculty shared design lore with their students by telling stories. In the mid-1980s, many of these stories had a common moral: the danger of using computers in structural analysis. One professor described a set of projects where young engineers used complex programs "not only without judgment but sometimes also without even analyzing the program's possibilities," producing "results [that] really are garbage." In a preemptive move, he insisted his students learn to program, but his effort had unhappy results. Those students who saw themselves primarily as designers thought programming was a waste of time. Others became so involved with programming that they drifted away from engineering. They spent more time debugging software than working on design.

This is a complex vignette. It points toward a future that would link the destinies of engineers and designers, even as many designers would be content to navigate the surface of their software, calling in technical "experts" when something went wrong.

Design is a volatile combination of the aesthetic and the technical. In the early 1980s, design students felt pressured to learn more about computers, but worried that this new mastery would turn their professional identities away from art and toward engineering. They needed to be computer fluent; they wanted to keep their distance. One way to accommodate the competing claims of fluency and distance was to form designer/hacker couples that divided the artistic and technical labor. The hacker members of these couples saw themselves as the midwives of the design future. The designers felt lucky to have found a strategy that allowed them to exploit simulation at a distance. These new collaborations prefigured how an architect such as Frank Gehry could later note that his designs were dependent on the most complex simulations, but that he himself did not get close to the computer.[8] More generally, they prefigured what would become commonplace in design offices—a senior designer, a principal of the firm, working on plans with a computationally sophisticated colleague at his or her side.

The simulations of the 1980s were fragile: programs crashed, data were lost. It took a lot of time to get anything done. These technical troubles obscured the reality that month-by-month, year-by-year, design software was overcoming its doubters. With some resistance, students were learning to live simulation as a new reality. It was a reality that subverted traditional hierarchies. As one civil engineering professor put it, in the new world order of design, the freshmen and sophomores were the "wizards" who could "run circles around the older kids." The faculty, once brilliant and revered, were "way behind, out of this." Faculty would have to "climb on board as best they can," once the field had been "turned around."

In classes, it was often graduate student teaching assistants, not faculty, who pressed students to learn the new technology, or as one undergraduate said, "If she [the graduate student teaching assistant] hadn't pushed, I don't know that we would have used it." Some teaching assistants became well known for their vigilance in pointing out bugs in the technical systems. One famously reminded undergraduates that "in industry you really need to use hand calculations to check the computer results."

THE SURPRISE OF STYLES

One of the hallmarks of the Athena project was that faculty were asked to build their own educational software. Most worked with a set of assumptions about learning: students begin by learning "fundamental concepts"; formal, mathematical representation is always the best approach; students would use simulations the way designers intended. As Athena unfolded, none of these assumptions proved true. To begin with, students approached simulation with a wide range of personal intellectual styles. Some used the highly organized style of the top-down planner, what many think of as the canonical engineer's style. These students spoke about the importance of beginning any project with the big picture. One student, working with AutoCAD, a computer-aided design program, explained how he used it. He said that he needed to start with "a diagram that gives you direction, a framework for your decisions. Then you can zoom in onto the details, work on them to reinforce the bigger idea, which you always have to keep in mind."

But other students found that, contrary to faculty expectations, AutoCAD in architecture and Growltiger in civil engineering were good environments for working in a bottom-up style, with little initial "framing." From their point of view, the programs let them take a design element and play with it, letting one idea lead them to the next. In simulation, design did not require precise plans; solutions could be "sculpted."

A graduate student in architecture was surprised that she could use the computer to work in this style, one I have called "soft mastery" or "tinkering."[9] When working this way, she approached design as an exercise "where I randomly . . . digitize, move, copy, erase the elements—columns, walls, and levels—without thinking of it as a building, but rather as a sculpture . . . and then take a fragment and work on it in more detail." This student saw design as a conversation rather than a monologue. She took the simulation as a design partner, an "other" that helped her shape ideas. She enjoyed the fact that the program seemed to "push back": "It pushed you to play," she said. In the same spirit, another student remarked that when he used the computer to sketch, he saw himself as a passive observer of spontaneous developments. This sense of the simulation as an "other" with a life of its own made it easier for him to edit, to "discover the structure within."

But even as simulation offered the opportunity to tinker and play, it could also encourage premature closure. As one student put it, "Even if you are just starting out, when you are faced with a computer printout, it is easy to feel as though the design has already been completed." Another described being "mesmerized by the hard-line quality and definition of the output and not want[ing] to change anything. And even if you do [make] changes, you do it in chunks, rather than making little moves, fine tuning, the kind of thing that comes from overlaying the tracing paper on an existing drawing and working on it." While some emphasized how simulation facilitated flexibility, this student focused on the deliberateness of every gesture in simulation. At least for her, "Everything you do is a very definite thing. You can't partially erase things, smudge them, make them imprecise." From the beginning it was dramatic: simulation offered different things to different people.

In the mid-1980s, what were designers thinking about when they were thinking about simulation? They were thinking about programming and design, doubting and doing, about attachment to the real through tracing with the hand. Having discovered the pleasures of tinkering with simulations, they wondered how wrong their first assumptions about computers in design might be. History had presented computers to them as instruments for "top-down" calculations; now they appeared as facilitators of "bottom-up" investigations. Designers had anticipated that computers would speed things up, freeing them from tedious calculations. What they had not fully anticipated were new ways of seeing sites and structures as they stepped through the looking glass.

INSECURITY

For architects, drawing was the sacred space that needed to be protected from the computer. The transparency of drawing gave designers confidence that they could retrace their steps and defend their decisions. In the 1980s, apart from all aesthetic considerations, the prospect of having to give up drawing made designers feel insecure. What of scientists? As with designers, scientists' insecurity about simulation was tied to concerns about opacity.

At MIT, chemists and physicists complained that when students used computers they entered data and did calculations without understanding their meaning. One chemistry professor said that even the most basic data-analysis programs led students to "simply enter a set of numbers, getting out an analysis, a number for which they had no clue if it was reasonable or not. They didn't even know whether they were working with a straight line or a curve." Most troubling, his students didn't seem to care. In response, he frantically rewrote computer code so that students would have to specify all analytic steps for any program they used. But even as this chemist worked to reintroduce transparency in software tools, his department was moving in the other direction. It began to use a computer program called Peakfinder that automatically analyzed the molecular structure of compounds, something that students had previously explored through painstaking hours at the spectrometer.

Peakfinder was a time saver, but using it meant that analytic processes once transparent were now *black boxed*, the term engineers use to describe something that is no longer open to understanding. When students used Peakfinder, they didn't know how the program worked but simply read results from a screen. One resentful student summed up the Peakfinder experience by saying, "A monkey could do this." Another likened it to a cookbook, "I simply follow the recipes without thinking."

Most of the chemistry faculty divided the work of science into the routine and the exciting and there was consensus that Peakfinder belonged to the humdrum. One said: "You see the same thing that you would on graph paper, but you're able to manipulate it more easily." From this perspective, simulation simply presented old wine in new bottles. But students could see what many faculty did not: computational speed and accuracy were quantitative changes that had a dramatic qualitative effect. With Peakfinder, the science on the screen began to feel more compelling than any representations that had come before.

So even as Peakfinder took chemistry students away from what many MIT scientists called "the messiness of the real," students said that being able to manipulate data on the screen made working with computer-mediated molecules feel more hands-on than working with "wet ones." Software made it easier to see patterns in data, to feel closer to what some students referred to as "the real science." They said that not having to worry about the mechanics of analyzing compounds with graph paper and plastic printer's rulers left more time for "thinking about fundamentals." Students noted that Peakfinder opened up chemistry to visual intuition. For one chemistry undergraduate: "There was the understanding that comes when you see things actually happen. The lines on the spectral graph were like seeing the molecule moving." For another, "We've always been told how molecules are moving, but it was the first time we actually saw what happens." These students' use of the word *actually* is telling. From the earliest days of Athena we have seen the paradox that simulation often made people feel most in touch with the real.

All science students have limited laboratory time and this makes it hard to collect good data. Before the computer entered the chemistry laboratory, if a student's one run of an experiment yielded only anomalous data, the student could not bring the result of an experiment into relationship with theory. To make this connection, the student was then given prepackaged data, data "certified" as correct. But Peakfinder saved so much laboratory time that students could afford many runs of their experiments. They were no longer so dependent on pre-packaged data. So, in addition to providing compelling visuals that illuminated theory, Peakfinder made students more likely to be collecting their own data, giving them a feeling of ownership over their experiments.

In physics, too, where computers were used to relieve the tedium of data collection and plotting, relatively mundane applications had significant effects. When calculation was automated and its results instantaneously translated into screen visualizations, patterns in data became more apparent. Physics students described feeling "closer to science" and "closer to theory" when their laboratory classes began to use software for visualization and analysis. As in chemistry, messy data no longer spoiled an experiment because you could afford the time to make extra passes until you had a good data set. One student was sure that doing all of these extra passes was what got her "hooked on physics" because "seeing that the data fits in spite of the variation is part of the allure of physics." Another physics student reflected, "If you do the same things 1,000 times by hand, you lose the sense of what you're doing. It takes so long, you forget the goal. You lose the forest in the trees." Another commented that for him "theory came alive" when a simulation demonstrated time and voltage in relation to each another. "You're not just seeing the curve drawn, you're seeing it actually happen . . . created point by point."

These students were experiencing another of computation's "paradoxical effects." Computers, usually associated with precision and rules, brought students closer to what is messy and irregular about nature. Computers made it possible for students to confront anomalous data with confidence. With multiple iterations, patterns became clear. As one student put it, "irregularities could be embraced."

Yet even with such benefits, and these benefits were substantial, MIT physics and chemistry faculty fretted that computer visualization put their students at an unacceptable remove from the real.

When students claimed to be "seeing it *actually* happen" on a screen, their teachers were upset by how a representation had taken on unjustified authority. Faculty began conversations by acknowledging that in any experiment, one only sees nature through an apparatus, but here, there were additional dangers: the users of this apparatus did not understand its inner workings and indeed, visualization software was designed to give the impression that it offered a direct window onto nature.

When chemistry students talked about feeling close to the "real molecule," faculty were at pains to point out the many, many levels of programming that stood between students and what they were taking as "reality." The most seductive visualizations could be described as a proverbial house of cards. For example, in chemical simulations, molecules were built from computer code that students never examined. Simulations of chemical reactions were built on top of these simulations, which were built on top of simulations of atomic particles. Professors tried to respect student enthusiasms while objecting to any uncritical immersion in the screen world. Many faculty began to argue the strategic importance of there being a place where they felt more in control of their students' development: this was the lecture hall. As drawing was for designers, as the analysis of structures was for civil engineers, the lecture hall became a sacred space for chemists.

In the lecture hall, chemists were confident that they could communicate what they considered the nonnegotiables. These included both things to know and how to know them. In their view, a computer could never teach theory, no matter what its usefulness in the laboratory. And a computer could never teach how science and the body are enmeshed. In the lecture hall, alone with their students, chemists felt they could do these things. In elaborate demonstrations, playing with models, turning and twisting their limbs to dramatize molecular reactions, chemists showed students how to think aloud and how to think with the body. And, in the lecture hall, chemists felt sure that they could show students how to put first things first. It was a place where faculty presented themselves as curators of their fields. There was similar sentiment among physicists. Lectures were a place where one could teach students how to be scientists. Although mentorship happened in laboratories, faculty acknowledged that there they would increasingly be competing with the seduction of screens. In lectures, one could hope for students' undivided attention.[10]

REVERENCE FOR THE REAL

At MIT, physicists became voices of resistance to Athena. When asked to be part of the project, their department several times declined. At the heart of its objection was the feeling that simulation stood in the way of the most direct experience of nature. Physics faculty acknowledged that some students found computers helpful for seeing patterns in the laboratory but then went on to speak reverently of the power of direct experience in their own introductions to science, or as one put it, "of learning Newton's laws by playing baseball." "Simulations," he said, "are not the real world. . . . There is no substitute for knowing what a kilogram feels like or knowing what a centimeter is or a one-meter beach ball. But these things have to be taught to students, and we faculty should teach it to them." For physicists, using simulation when you could be in direct touch with the physical world was close to blasphemy.

Physics faculty were concerned that students who understood the theoretical difference between representation and reality lost that clarity when faced with compelling screen graphics. The possibility for such blurring existed even when students used simple programs that visualized data. Indeed, the faculty who introduced the first generation of such programs were shocked at how little it took for students to get lost in what would come to be known as "the zone." It was a time when a simple computer game such as *Space Invaders* mesmerized millions. As simulations became more complex, faculty anxiety grew. For one physicist:

My students know more and more about computer reality, but less and less about the real world. And they no longer even really know about computer reality because the simulations have become so complex that people don't build them any more. They just buy them and can't get beneath the surface. If the assumptions behind some simulation were flawed, my students wouldn't even know where or how to look for the problem. So I'm afraid that where we are going here is towards *Physics: The Movie.*

In physics, student opinion mirrored faculty concerns. For one physics student, "Using computers as a black box isn't right. For scientists who are interested in understanding phenomena and theorizing, it's important to know what the program is doing. You can't just use it [a program] to measure everything." Another student admitted that it was easy to "be mindless on the computer. You put worthless data in and get worthless results out, but think it's noble somehow just because it has been through the machine." Although some students spoke about how automatic data collection enabled them to see new patterns, others extolled the virtues of taking data by hand, even though this might mean getting only a fraction of the data points. For one, "Doing it by hand forces you to think about the data as it is being taken. You have to know what you are looking for." In his view, working through a long series of hand calculations put him in a position to keep track of the kinds of errors he made. His dedication to the aesthetic of transparency was total: "When you work on the computer it is hard to tell what data is 'lose-able' and what is essential. The computer makes these choices for you. That's why I do it by hand as much as I can. Doing it by hand, point by point, you can see subtle variations in the phenomenon you are studying." Another student described "chart[ing] tables by hand" as providing "an intuitive sense of an experiment." One "live[s] with these numbers so much that you begin to understand what they mean. It's nice to be able to enjoy your graph and take your ruler to it. With the computer, it's a mental process further removed from reality. The problem is in getting the feeling of moving the ruler on the paper with the computer."

Just as this physics student spoke about the pleasures of ruler and paper, another admitted that even though doing things by hand was "drudgery," it was useful drudgery:

There is some truth to the importance of getting a feel for the data by hand. When you plot it with the computer, you just see it go "ZUK" and there it is. You have to look at it and think about it for a long time before you know what it all means. Whereas when you draw it, you live with it and you think about it as you go along.

No member of the MIT physics faculty expressed greater commitment to transparency and the direct experience of nature than William Malven. Of direct experience he said, "I like physical objects that I touch, smell, bite into. The idea of making a simulation . . . excuse me, but that's like masturbation." On the need for transparency, Malven was a purist. Ideally, for Malven, every piece of equipment that a student finds in a laboratory "should be simple enough . . . [to] open . . . and see what's inside." In his view, students should be able to design and build their own equipment. If they don't do so for practical reasons, they should feel that they could have. Malven saw his role as a teacher and a scientist as "fighting against the black box." He admitted that this required a continual "rear-guard action" because, as he put it, "as techniques become established, they naturally become black boxes." But, he added, "it's worth fighting at every stage, because wherever you are in the process there is a lot to be learned."

Yet even as Malven understood simulation as a potentially threatening black box, he was dedicated to the idea that scientists could create visualization programs that would improve scientific training. The key was for scientists to take control of their own pedagogy. For example, Malven envisaged software that would teach transparent understanding by requiring researchers to make explicit the procedures they were demanding of the computer. To this end, in the early 1980s Malven designed a laboratory course for all juniors majoring in physics. He wrote all of the course software himself. All of his programs, designed to help with data collection and analysis, had their inner workings transparent to the user.

Malven's junior laboratory began with a classic experiment that used computers to increase students' sensitivity to experimental error. When a source of electrons is placed first in a magnetic and then in an electric field, it is possible to determine its specific charge. Malven set up this experiment using a computer program that would not accept a data point without a student specifying an error factor. Malven described this as an effort to use simulation to bring students closer to "the real data."

A second experiment, known as the Stern-Gerlach experiment, used computers to similar effect.[11] Here, a beam of silver atoms is passed through a magnetic field and then onto a film plate. The atoms form smudges from which their orientation can be determined. Originally in a mixed state, the silver atoms collapse into one of two orientations—as quantum physics would predict—once they enter the magnetic field.

Before the computer was used to collect data in the Stern-Gerlach experiment, students had to manually move magnets, read a meter, and take measurements with a painfully slow pen-recording device. Malven said that in those precomputer days, "you could never quite figure out what was going on because it would take fifteen minutes to do any one thing." But Malven devised a new way of doing the experiment: he attached a computer to an apparatus that collected data from the film smear and he wrote software that did the calculations necessary to determine the emergent shape. Because of the computer's speed, two peaks appeared almost instantaneously, indicating the two magnetic moments. For Malven, the dramatic appearance of the two peaks made the experiment come alive, an example of "something profound and something mundane—that combination, it's like love! The mundane part: something as simple as remembering to degauss the magnet in the experiment. The profound part: the space quantization. That the mundane and profound go together—like washing dishes and love—it's one of those things that is hard to learn."[12]

Malven said that using the computer brought his students closer to science because they could play with data and tinker with variables. Like Peakfinder in chemistry, Growltiger in civil engineering, and computer-aided design programs in architecture, the software in the junior physics lab encouraged students, in Malven's words, to "ask a question and say, 'Let's try it.'"

The instrument that Malven used most in the laboratory was his Swiss Army knife, a simple, all-purpose tool. He thought that the best computational environments were—like the knife—transparent and general purpose. He was not happy when the physics department purchased some "fancy data analyzers." In his view, only general-purpose computers gave students sufficient opportunity to write their own programs, to learn from "the ground up." For Malven, knowing how to program was crucial to making simulations transparent. Other people wanted to use the most up-to-date tools. Malven was interested in only the most transparent ones.

Malven's involvement won over some of his skeptical colleagues to the idea of using computers in their classes under the banner of transparent understanding. Physics professors Barry Niloff and David Gorham designed an Athena-sponsored freshman seminar that began by having all students learn to program. Then, the seminar turned to computers to engage students in problem solving that used numerical rather than analytical methods. Professor Niloff, summed up the seminar's purpose—to demonstrate the power of "just plotting stuff," using the computer to get students closer to direct observation. Only a small subset of real-world physics problems are solvable by purely analytical methods. Most require experimentation, where you do trials, evaluate forces, and fit data to a curve. The computer makes such numerical solutions easier to do. And in a practical sense, it makes many of them possible for the first time.[13]

Over time, the freshman seminar turned its focus to problems of physical measurement. It was something about which Niloff and

Gorham were passionate. Gorham noted that from the time students began to use calculators, they suffered from an insufficient understanding of scale and of "what it means to make a physical measurement," in particular, what it means to have a margin of error. Students with calculators had gotten lazy; they didn't want to do things by hand, and they didn't want to include units in their calculations. The stakes were high. As Gorham put it, for a physics student, not understanding error during college could translate into a "space shuttle blowing up" after the student entered the workforce.

Gorham believed that computers were making students' problems with scale and error worse but that since calculators and computers were not going away, good pedagogy demanded that students be forced to do "back of the envelope calculations." These calculations require an understanding of the scale you are working in, the units you are using, the number of significant digits that make sense. Gorham saw himself as on a mission: to take a technology that had caused problems and turn it to the good. Computers had played a part in encouraging students to be sloppy about scale and error. But now teachers could use computers to become more effective "proselytizers for statements of error terms."

In one freshman seminar assignment, Niloff and Gorham set up a computer program that simulated a ball dropping in space. Students could vary the ball's weight and the height from which it was dropped. The program displayed the falling ball, recorded relevant data, and then allowed students to analyze it. Students were asked to consider the influence of wind drag and inertia on the measurements. Gorham said:

In the ball dropping exercise, the problem was simply stated. Measure the acceleration due to gravity and then state the uncertainty. And this blows their [the students'] minds totally. People are used to textbooks. They are used to computers. They don't understand that the computer is marvelous to take data and reduce it, but what does it mean to have an error? If someone said, "g is 9.81," that's totally meaningless. 9.81 plus or minus what? 9.81 plus or minus 10 is not a very good measurement. And so we try to drill into them the whole idea of error analysis.

Like Malven, Niloff and Gorham believed that students needed a deep knowledge of laboratory software that could only come from knowing how to program it. Niloff went beyond this: students needed to understand how computers worked, down to the physics of the processor and the graphics screen. Understanding these deeper levels would ultimately bring students closer to problems of estimation, scale, and error. Said Niloff, "When students plot points for the first time, they literally understand what the physical screen is, what the graphics screen is, how to actually put points on the screen." A curve drawn on a screen and a theoretical curve might look the same, but a student who understood the screen's resolution might find a difference at "the tenth of a pixel level, which you can't see. . . . And this tenth of a pixel level may be of critical importance. It may be the margin of error that makes all the difference."

SIMULATION AND DEMONSTRATION: SCIENCE AND ENGINEERING

The junior laboratory and freshman seminar were notable exceptions to a general "antisimulation" culture in the physics department. But in one instance, even those faculty most hostile to simulation were willing to accept it as a necessary evil. This was when simulation made the invisible visible—that is, when it provided access to quantum-level phenomena.

In the 1960s, MIT physicists Harold Rabb and Burt Fallon were part of a research group whose focus was innovation in physics education. Rabb and Fallon continued this work and, among other things, collaborated on short films that illustrated principles of relativity. One of these films demonstrated wave packets propagating as a function of time and fragmenting upon collision. It made quantum mechanics come alive. During the 1980s, Fallon used interactive computing to continue his visualization work in the areas of relativity and quantum physics. He wanted to go beyond the passive presentations possible in film to provide an experience of *living* in the quantum world, including the ability to perform experiments within it. Fallon's computer programs were designed to demonstrate the invisible physics, to help students develop intuitions about the quantum world the way they developed intuitions about classical physics: by manipulating its materials.

One of Fallon's programs simulated what it looks like to travel down a road at nearly the speed of light. Shapes are distorted; they twist and writhe. Objects change color and intensity. All of this can be described through the laws of physics, but Fallon pointed out that "you can't experience them directly, except through the computer." When MIT physicists in the 1980s looked at Fallon's work, they saw it as an exception to a cherished rule. They were willing to accept a simulation when no real-world experience could possibly be substituted, but when Fallon used simulation to demonstrate something that could be done in a traditional laboratory setting, one saw the full force of his colleagues' hostility.

The following exchange on simulation between physicist Barry Niloff and his colleague, Arthur Richman, spoke directly to this issue. Here Niloff and Richman are eloquent about the importance of keeping simulation in its place.

Richman: One of the problems that a physicist has to come to grips with is that sometimes light behaves like a particle and sometimes it behaves like a wave. If you have a dike and two little openings, the waves of water will propagate through those two little openings. They'll form little rings, which will then interfere with one another, and you'll see the results of these two wave fronts coming out, interfering with one another. That's a very clear wave front. If you think about shooting bullets through these two holes, you know the bullet goes through one or through the other. Now, you're being told as a student of quantum mechanics that sometimes you're supposed to think about light in one way and sometimes you're supposed to think about it the other way. And so a very important experiment comes to mind. You take this case of two slits and decrease the level of illumination so you're very sure photons are only going through one at a time. You would be tempted to say, "Well, by gosh, this is going through one at a time. It's like a bullet, it goes through there, or it goes through there." And there's a dramatic demonstration that can be done to show in a way that just hits you over the head in a beautiful way, that even though they're going through one at a time, they are managing to diffuse. It's a fantastic experience for a physicist who is beginning to think about quantum mechanics. And I think many of us have the same reaction, that to simulate that on the computer . . .
Niloff: It's a cheat.
Richman: It's almost sacrilege.

William Malven worried that students, drawn to substituting models for reality, are then tempted to believe that something will happen in the real world because they have seen it in a simulation. Richman and Niloff feared that when their colleague Fallon demonstrated something on the computer that could be shown without it, students would have a feeling of understanding without true insight. On the most general level, in the 1980s, MIT physics faculty opposed anything that smacked of demonstration rather than experiment. A demonstration takes what we know and shows it in powerful relief. An experiment, in ideal terms, turns to nature ready to be surprised. But if experiments are done "in simulation," then by definition, nature is presumed to be "known in advance," for nature would need to be embedded in the program.

Physicists acknowledged that, in their own way, simulations can surprise. Complexity in simulation can lead to "emergent effects."[14] But in simulated experiments, they insisted, the programmer has always been there before. Said one student, commenting on an experiment presented virtually, "The 'experiment' is prewired," and it was this prewiring that had turned it into a demonstration, or as Richman summed it up, "close to sacrilege." Physicists constructed nature as untamed and unruly. Simulation sorts it out.[15] At MIT the debate about simulation and demonstration was fueled by faculty anxiety that if students performed enough "experiments" in simulation, they would become accustomed to looking for nature in representations they did not fully understand.

Simply put, for MIT physicists in the 1980s, simulation provoked discontents. Physicists used criticism of simulation as a way of asserting core values defined in opposition to it: the importance of transparent understanding, direct experience, and a clear distinction between science and engineering. For them, simulation was dangerously close to demonstration, the stuff of engineering education. They conceded that engineers might be satisfied with simulation, but as scientists they were bound to resist it as something that might taint scientific culture with engineering values. And indeed, physics faculty felt that they could already see signs that simulation was eroding critical sensibilities. For one thing, when students found something on the computer, they tended to assume that someone in authority had thought it was correct. For another, students who had grown up with video games experienced screen objects as, if not real, then real enough. This was a generation more likely than their elders to take the screen world "at interface value."[16]

If physicists worried that the computer was a Trojan horse that might introduce engineering values into the scientific enterprise, they were also concerned that it would make physicists too dependent on engineers. One student put it bluntly: "Of course you can't know everything about the computer—as a computer science major might—but you should know enough that you don't have to hand your work over to a Course 6 [Electrical Engineering and Computer Science] guy."

Another physics student said he respected engineers but drew a sharp distinction between his field and theirs: "A goal of engineering is to create new devices of human significance." In physics, however, "one is dealing with the universe. It elevates you to an eternal sense . . . in physics you're relating to absolute truth as opposed to practical truth." In his view, simulation did not help with absolute truth.

Engineers were rather less conflicted. They contented themselves with personalizing software (even if some elements of it were black boxed) by tweaking it in a style MIT students called *customization*. In the 1980s, customization referred to making opaque software do something quixotic, often "against its grain." So, for example, civil engineering students took a program that was supposed to drill them with questions and got it to do their homework for them.

Although some joked that they didn't have a "physicist's" understanding of the simulations they modified, this kind of playful interaction gave a feeling of making technology one's own. Customizing was a strategy, as one civil engineering student described it, "to feel less at a program's mercy than at its command."

Physicists were critical of customization. In their view, there was no true intellectual ownership without full transparent understanding. For them, customization exemplified the kind of compromise that engineers and designers might be willing to make, but that a scientist should never consider.

In the 1980s, scientists were comforted by the idea that building simulations was the province of computer scientists and that they were in a very different field. Not too much time would pass before this division became elusive. More and more, science was done on the computer and a migration of computer scientists to the natural sciences would soon make it hard to say where one left off and the other began.

DESIGN AND SCIENCE AT THE MILLENNIUM

In the 1980s at MIT, many early users of simulation genuinely could not imagine such things as designing without drawing or thinking without a "back of the envelope" calculation. Those who had grown up accustomed to physically taking apart their laboratory instruments were upset by programs whose inner workings they did not understand. In response to simulation's provocations, faculty and students identified areas that they hoped to keep as simulation-free zones. Architects wanted to protect drawing, which they saw as central to the artistry and ownership of design. Civil engineers wanted to keep software away from the analysis of structure; they worried it might blind engineers to crucial sources of error and uncertainty. Physicists were passionate about the distinction between experiment and demonstration. They believed that computers did have their place in the laboratory, but only if scientists were fluent with the details of their programming. Chemists and physicists wanted to protect the teaching of theory—the elegant, analytical, and inspiring lectures of great MIT scientists were the stuff of legend.

The 1980s were marked by substantive disagreements about the role of simulation and visualization in science, engineering, and design. These days, the space for this kind of disagreement has largely closed down; in the past twenty years, researchers have gone from using simulations for discrete, tactical purposes to working almost full time in simulation.[1] Over time, factions for and against the computer have been replaced by individuals expressing ambivalence about what has been gained and lost. Protecting sacred spaces has given way to enduring anxieties about life on the screen.

Generational differences influence the distribution of this anxiety. An older generation feels compromised by simulations that are essentially "black boxes"; using them seems an abdication of professional responsibility. A younger generation is more likely to accept that computational transparency, in the sense that their elders speak of it, is a thing of the past. Indeed, today's professionals have watched the meaning of the word *transparency* change in their lifetime. In the early days of personal computing, command lines on a screen reminded users of the programs that lay beneath. With the Macintosh in 1984, users activated screen icons with a "double click." Transparency once meant being able to "open the hood" to see how things worked. Now, with the Macintosh meaning of transparency dominant in the computer culture, it means quite the opposite: being able to use a program *without* knowing how it works. An older generation, one might say, is trying to get a younger to value experiences they never had and understand a language they never spoke.

Recall the MIT faculty member who feared that students could not use simulation and maintain critical distance from it at the same time. He thought that the problem of "doing and doubting" concerned novices. Time has proved otherwise: simulation seduces even experienced users. For one thing, these days the body is routinely brought into simulation—think of chemists who manipulate screen images of molecules with gestures they once used to twist physical models. When the body is part of the experience of simulation, doubting is difficult even for experts, because doubting simulation starts to feel like doubting one's own senses.

Today, those who grew up in the days of Athena hold positions of professional authority. Like their teachers, many see the limits of simulation, but they face different challenges than the generation that came before. With research and design now indissociable from simulation, one cannot simply put a pencil back in the hands of a designer or ask a molecular biologist to model proteins with balls and sticks.

But even if the notion of sacred space now seems quaint, what remains timely is finding ways to work with simulation yet be accountable to nature. This is a complex undertaking: as we put ever-greater value on what we do and make in simulation, we are left with the task of revaluing the real.

NEW IDENTITIES

Although many architects and planners in the 1980s looked toward a future when designers would be computer-fluent, most continued to define their professional identity by contrasting themselves with so-called "computer types." Two decades later, a basic tension remains. Even as computer-aided design has become common-place, it is just as commonplace for design professionals to describe who they are by making clear what they do *not* do with computers.

In a Spring 2005 MIT workshop on simulation and visualization, an MIT architecture professor trained during the Athena years con-trasts designers and technologists: "I'm absolutely skeptical. Can those two mentalities exist in the same brain? I haven't met the per-son yet who is a designer and a programmer." An MIT student at the workshop concurs by distinguishing between design logic and computer logic, complaining that the codification intrinsic to com-puter logic inhibits his creative thinking.

The resistance of individuals to simulation shows up in the social world of design firms. Instead of the heated debate of the 1980s, these days one sees more passive strategies: not showing up for meetings, learning computer skills and choosing not to use them, demanding to use old techniques next to the new, launching com-plex negotiations about when designs should be digitized.

In one practice, an architect in his thirties turns to the colleague who will teach him how to use a design tool known as CATIA (Computer-Aided Three-Dimensional Interactive Application) and says: "Why do we have to change? We've been building buildings for years without CATIA."[2] His instructor, an engineer who has intro-duced CATIA to several architectural practices, is familiar with this kind of comment; he ruefully characterizes three difficult phases of resistance to his teaching. The first is "the brick wall": architects say they are too busy to learn. They argue that learning how to use digital technology is time consuming, so much so that it will exclude other kinds of practice. Then comes a "tutelage with resistance" phase, when the firm's principal designer insists that his architects learn the program. In a final, "implementation with continuing resistance" phase, firm architects finally use the program but find cause for constant complaint. Some argue that CATIA helps consultants and contractors but not designers. Some complain that it produces drawings that are "cluttered, both visually and conceptually."

The CATIA instructor sees computer-aided design as a new way of looking at the world while his colleagues tend to describe it as "just a tool." As in the 1980s, the phrase "just a tool" is charged with the work of keeping the computer in its place, away from the core of architectural identity. But these days it is less common for designers to reject simulation technology than to accommodate it and complain about the problems it fosters.

Marshall Tomlin, a young designer at the firm where CATIA is being introduced, laments that much of his work, rendering architectural drawings, consists of choosing among options on a computer menu. He admits that he is always tempted to go with the "default," the choice that the system offers unless you specifically choose another. He wishes that his work felt more "his," but a sense of authorship eludes him. And he worries that his drawings mislead. He explains that when rendering was done by hand, detailed drawings signaled a commitment to a design program. Now, he adds details to what look like final drawings while his firm's engineers are still working to create the underlying geometry of the plans. Design firms have always used beautiful drawings to sell not-yet-completed projects. For Tomlin what has changed is that computer drawings make all buildings look as though they have been fully considered, designed down to the last detail.

Beyond issues of authorship and his anxiety that his work creates an illusion of commitment, Tomlin thinks that the use of computers in his firm leads to a greater rigidity of roles, an increased tendency to identify people by their function. His job as a renderer has come to feel reduced to a particular relationship with the computer. He is not happy that his colleagues seem content to leave him alone with his machine.

Tomlin's firm uses both computer-aided design systems (CAD) and technologies (CAD/CAM) that support design, project management, and manufacturing. One group of architects uses the computer to sketch out the basics; another group enters the design on a computerized system that sees it through production. Most of the firm's architects don't know much about the workings of this second system. In the past, Tomlin and his colleagues sent off preliminary designs to be specified in more detailed drawings and cardboard models. They say that those physical drawings and models remained accessible to them, open to being fine-tuned by the organization as a whole. These days, when they hand over their designs, they feel loss. One of Tomlin's colleagues says that when he sends a design to the technical group he immediately feels left out: "It's going into the land of the hackers, people who don't necessarily know design the way I do."

Often, when designers seem to be objecting to a particular computer system, they are really objecting to how the machine forces them to abdicate control over their design. In most firms, there is social pressure to do everything and put everything on the computer.[3] It is the gold standard for current "best practices." In response, designers make efforts to balance their experience. They sit in front of giant screens, but on their desks are plastic building blocks, clay, wooden dowels, cardboard, and glue. One young architect explains how hand sketches and cardboard models "preserve my physical intuitions." Another, who describes herself as an "AutoCAD baby," says that while the computer enables her to go "inside the materials," she still needs physical models to recall what she terms their "atmospherics." Designers check their computer models against cardboard ones that they describe as more "real." They wait until their designs have fully stabilized before bringing in digital tools, even if the intended purpose of these tools was to afford the opportunity to play with design ideas-in-process. One says he lies down on computer printouts, bringing his body into the world of the simulation.

In Tomlin's firm, the principal designer works with computer design tools by sitting next to a technologically adept apprentice.[4] When the tools were first introduced, the master architect made sketches that his apprentice translated into a geometric model on the computer. The master made revisions by working with tracing paper over the model's printout. Over time, the master stopped requesting printouts. He began to make changes directly at the computer, always in the company of his apprentice. The master does not work alone but has found a way to stay close to the evolving design.

Some senior architects welcome a new alliance, indeed a new symbiosis, with what several call a "digital person." Others feel uncomfortable with this kind of dependency, either rejecting any technical involvement or insisting that they have to master the technology themselves. They try and succeed or they try and fail and make plans to try again. It is hard to have both the responsibilities of being a senior designer and enough time to learn complex computer systems.

At the Spring 2005 MIT workshop on simulation and visualization, the architect Donna Gordon calls herself a "digital person" and speaks to the complexity of such partnerships. She herself has worked with several master architects whom she describes as people "looking over her shoulder." While it is the master who directs the action, Gordon feels that it is she and those in her position who are in the more intimate relationship with the design. She suspects that from the perspective of the master, the technical apprentices are merely "cranking it out." But from her point of view, the technical apprentices are better able to see when "something is wrong or something could be better. . . . They are the ones who are so intimately focused in on it. They are three-dimensionally seeing the space from inside-out." It is the technical apprentices who are "sculpting space."

As she works, Gordon feels herself "falling into the model," developing something she experiences as a body knowledge of its contours. Her job in working with master architects is to get them thinking virtually—to get the mentor inside the model. Her strategy is to take the master architect on a walk-through of the building, renamed a "fly-through" when one speaks of digital space. During the fly-through, Gordon rotates the model to reveal hidden structures; she zooms in and out to give the master architect a kind of guided tour.

The digital model is not simply shown, it is performed.[5] In the process, observers are brought into a new relationship with what is on the screen. Gordon describes this as "bringing that person into the intimate connection . . . you are taking them in by the hand." She has only good things to say about her own experience of that intimate connection. But sometimes, architects can "fall into a model" and have a hard time getting up, in the sense of maintaining a sight line outside the simulation. In the Athena experience, when a missing set of contour lines on the screen led a student away from the topography of an actual site, it was reasonable to blame the mistake on the limitations of a primitive computer system: it could not fit enough contour lines on the screen to present the site in all its detail. Today's sophisticated systems do not have that problem. In fact, now it is their fluid and detailed virtual realities that can sometimes edge out the real.[6]

EDGING OUT THE REAL

In the Spring 2005 MIT workshop, one young architect says that he has lost "references" outside his digital models. "It is always an interesting kind of breaking point," he says, "where the simulation is so novel that you can't judge it anymore. Because you don't have a reference to say that this is wrong or right. Because there is no frame of reference . . . no precedents." His digital tools are designed to capture the feeling of working with traditional materials, but he had no part in designing them and does not understand how they are made. Most distressing to him is that when he gets confused and loses his "frame of reference," his model can feel more compelling than any real building. Simulation mesmerizes.

His colleagues discuss an example on this point: a high visibility project that envisaged replacing a block of century-old brownstones with a luxury hotel in a traditional European style. To cut costs, developers asked for revisions to a planned design. The designers at the workshop imagined what had happened next: architects used a CAD/CAM program that allowed them to open a stylebook, click on a surface, and "paint" it on sections of the building. At the computer, the designers clicked on a limestone look. This translated into an order for simulated limestone, limestone surfacing etched on fiberglass. Additionally, a choice was made for cutout dormers, less expensive to build than real ones. On the screen the simulated limestone and the cutout dormers might have looked acceptable. But when built, the hotel provoked an outcry. Critics described it as worthy only of Disneyland. The physical building looked like a simulation,[7] out of place surrounded by "real" buildings. The hotel's developers were required to redo its façade, one that had been born of and, one might say, *for* simulation.

In the terms of Louis I. Kahn's question about bricks, we might ask "What does fiberglass want?" and find the answer is not French Second Empire architecture. Although news reporting of this architectural cause célèbre made it clear that the motivations for design changes were financial, at the MIT workshop, the story is discussed as a cautionary tale about the risks of making design decisions on the computer. The architects at the workshop speak about how easy it is to make mistakes when materials are chosen with a double click. One comments that for him, layers of simulation are like "levels of myth making." He says that when designing on the computer, "We believe that the space is going to become what we see." He calls this "the visualization/reality blur."[8]

When using a CAD/CAM system, it is the system that manages contracting and purchasing. The architect who specifies materials is at a remove from the craftspeople who will actually construct the building. The architect who worries about the "visualization/reality blur" recalls that craftspeople once understood "what the building was going to become"; they once "took part in the design process." CAD/CAM has disrupted these relationships. For him, craftspeople are no longer colleagues in the old way; workers with their "real" materials are less likely to be "building the thing that we [the designers] visualize."[9]

Designers have commented that there may be a connection between computer-aided designs that go awry and the sensibility encouraged by the digital fly-through where architects sweep through their simulated buildings, always in movement. The hotel with the faux-limestone facade was the kind of building that looks best either on the screen or in the blur from a highway where the observer is put in a situation of speed and movement that calls to mind the sweep of a fly-through.

Even architects who feel confident that *they* would never design such a problematic building are respectful of the seductions of simulation. They understand that there can be a day at work when fiberglass limestone and cutout dormers look good on a computer. Working in simulation, one often has a feeling of exhilaration, of being liberated from traditional materials. It is not unusual for this experience to be brought up short by the resistance of the real.

One young architect at the Spring 2005 MIT workshop says that graduate education taught her that architects must be "careful with visualizations." The gap between the building on the screen and on the ground can be huge: "Architects," she says, are "losing their connection to materiality." With irony she adds that when designers use CAD/CAM, it is too often the case that "the architect has left the building."

RECONSIDERING SACRED SPACE

In the 1980s architects tried to protect drawing from the inroads of computation. Indeed, they wondered if those who could not draw should be architects. In drawing, they argued, the architect felt as well as thought the building. Drawing was the place for inspiration. Over time, simulation has come to occupy (some would say, has encroached upon) this sacred space.[10] While some architects are comfortable with this development, pleased that a larger group of people are now able to participate in design, others remain skeptical.

One such skeptic, Howard Ramsen, now in his mid-fifties, went to design school in the 1970s. He was drawn to architecture by his love of art. The most pleasurable aspect of design, he says, was the time he spent sketching. But by the early 1990s, he forced himself to learn computer-aided methods to maintain his competitive edge in the profession. "But I never liked it," he said. "I didn't become an architect to sit in front of a computer." After a few years, he felt a conflict: "I love to draw, I think as I draw, that is how my ideas come to me. When I draw a building, I have confidence in the building. I know the building in a different way than when the computer draws it for me."

For Ramsen, the palettes, menus, and default programs of computer-aided design make him feel less like the author of his buildings. Looking back on his career, Ramsen says he felt most alienated when he worked in a large firm where designs were developed by offering architects a set of "starter elements." At the computer, Ramsen missed the flexibility of smudging a pencil line with his fingers. For him, the designs seemed to have moved "inside the computer," where he couldn't touch them; designing itself had become more like puzzle solving. At first, he comments on this experience by saying that "design had lost its artistic fluidity." But a few moments later, he phrases it otherwise: "I lost my artistic fluidity." When he worked with a machine that he experienced as "drawing for me," it had the effect of undermining his sense of authority:

When I draw a building myself I check all the dimensions. With the computer, well, I don't. It seems presumptuous to check, I mean, how could I do a better job than the computer? It can do things down to hundredths of an inch. But one time, on a big project, the computer drawings came back and I didn't check dimensions and the foundation was poured. We didn't know until the contractor started framing that there had been a mistake. All because I grew up to be intimidated by the authority of the printout.

When Ramsen draws a building, he feels a responsibility for it. He commits himself to what, in the 1980s, MIT professor Ted Randall called the designer's "marks." But when presented with computer output, Ramsen feels deferent, outmatched. How could he be more precise than something that makes discriminations "down to hundredths of an inch"? In the case of the project with miscalculated dimensions, the computer's precision caused Ramsen to confuse precision and accuracy; he assumed that his computer's output would not only be precise but correct. Reflecting on that project, Ramsen knows that at some point he made an input error, an error that was not the computer's doing. But despite himself, he had come to experience the computer as a kind of correction machine. As a result he no longer checked the drawings the machine produced. His contractor (who also did not check the computer drawings) confessed that when designs had been hand-drawn, he checked them all the time. But when given computer-generated drawings, he just went ahead and poured the foundation they specified. "Intellectually," says one of my students, "you spend your lifetime with computers learning that it is 'garbage in/garbage out.' But the fancier the computer system, the more you start to assume that it is correcting your errors, the more you start to believe that what comes out of the machine is just how it should be. It is just a visceral thing."

Ramsen ended up leaving the big firm and going into private practice. This is where he works at present; he likes paying close attention to small projects. He designs by hand and sends his drawings out to be "put into the computer" by somebody else. Hand drawing makes him feel closer to his roots in design. And he feels that his new distance from technology makes him a better architect. He says:

The computer makes all sorts of things possible, buildings that you could never, ever have built before. But just because you can build a building, doesn't mean you should build a building. So many buildings today are "extreme," they test the limits of what material can do. But they are really not solid. The computer said that everything was okay, but people can't understand the physics of buildings at that level of complexity, so a lot of mistakes happen.

In our conversation that had begun with his relationship to drawing, Ramsen moved from considerations about aesthetics to the fantasies he had developed about computational precision. He then turned to questions of process. Does the computer enhance flexibility in design or close it down?

Everyone says and you would think that the computer is supposed to put you in a state of mind where you try this and that and keep making changes because it is so easy. But in the firm where I worked, because it [the computer] presented everything at such a level of detail, the building seemed finished after we had put in pretty much our first idea. So, I found that things stopped being in process way too soon because everything looked so finished.

A pattern emerges: in simulation, architects feel an initial exhilaration because of the ease of multiple iterations. But at a certain point, the graphics are so spectacular, the sketches so precise, that possibilities can feel like inevitabilities. Detailed hand drawings once signaled that major design problems had been resolved; when computers produce such drawings, they cue a similar response. Today's designers may experience that sense of completion, even when they know it is not warranted. As the renderer Marshall Tomlin found to his distress, in digital format, even preliminary ideas look finished.

As we have seen, simulation produces paradoxical effects. Despite offering the possibility of multiple iterations, in simulation, it often turns out that the first idea wants to be the last idea. When confronted with a detailed computer-generated drawing, one could simply undo what has been done. But in practice, the fine resolution of screen drawing is more likely to persuade people to accept it as a fait accompli. Currently, this makes drawing not a sacred space but a contested one that preoccupies architects. As one designer puts it, "Everything we do when confronted with a new project is to figure out when we stop drawing, what we draw, what we need to draw, what we don't need to draw and don't want to draw, what we expect others to draw."[11]

So while some take the loss of hand drawing as the cost of doing business in contemporary practice, others feel that, without the sense of ownership that comes from the sweep of hand on paper, design is diminished. Today's architects face a beautiful screen. But it may be that the master architect we see leaning over the shoulder of the apprentice will still choose to walk through the fly-through in his mind's eye, asking: what does the real want?

LIFE SCIENCES: THE TENSION BETWEEN DOING AND DOUBTING

What does the real want in the life sciences? There, simulation began with a kind of deception, an aesthetic compact with nature. Early simulations were qualitative and evocative. For example, one biophysicist, Stéphane Leduc (1853–1922), working at the Nantes Medical School, drew on the persuasive powers of mimicry to simulate the mechanical processes governing life forms. He used salt crystals and dyes to produce artificial cells and organisms. These chemical creatures, formed by virtue of osmotic gradients, seemed strangely alive, their growth mimicking those of dividing cells, sporulating mushrooms, blooming plants, and free-swimming algae. Leduc's simulations mimicked life without reference to its underlying processes.[12]

As biology matured and computation became its dominant tool, this kind of simulation was discredited. These days, a life scientist at MIT, who models protein-protein interactions and trains a new generation of biological engineers, describes a model that has no mathematical precision or predictive capability as just a "cartoon."[13] For him, anything less than the quantification of physics at a molecular level is "mere philosophy." These days, visualization and simulation underpin biology as it manipulates and reengineers life at the molecular and cellular level.[14] Mathematical simulations animate models that represent proteins and cells over time. Algorithms predict molecular interactions within cells and the pathways of protein folding. Scientists have built a second nature within the computer through simulations that are ever more manipulable, ever more easily experimented on. Some describe the result of such virtual practices as "new forms of life."[15]

In today's biology, the simulation of life is central, but getting it right has remained elusive. Living systems work on many levels, from atoms to organisms; integrating levels is difficult. And the intricate workings of cells and molecules are hard to see, quantify, and analyze. These challenges encourage some life scientists to approach simulation with what the MIT architects of the 1980s termed a *critical stance*. In the life sciences, a critical stance toward simulation enforces modesty. In the field of protein crystallography, which uses X-rays to investigate molecular structure, some researchers take pains to insist that the models of complex molecules they produce are "just models." Microscopists are quick to describe the extent to which their images are only mediated representations of the cells they study. These life scientists take as a given that simulations can deceive and that to assess simulation one must find a vantage point outside of it. Simulation and visualization have become the everyday workplace of life sciences. But the programs that scientists use are typically "black boxed." In this way, scientists' feelings of mastery become tied to anxiety and uncertainty.

Computers first came into protein crystallography in the late 1940s. Their job was to lighten the labor that stood behind crystallographic calculations, labor that had typically been allocated to women, who ironically were known as "computors."[16] By 1957, computers had been used in the construction of the first visualization of a protein, a model built by hand out of Plasticene and wooden pegs. Computer graphics for molecular visualization came later, in the mid-1960s.[17] From this point on, scientists would build molecular models by interacting with computer graphics; physical models were too cumbersome.

These days, the X-ray diffraction analyses that protein crystallographers depend on are collected, measured, and calculated by computers. For the most part, protein crystallographers have welcomed this innovation. In 1959, it had taken Max Perutz and his team of technicians twenty-two years to complete his Nobel Prize–winning model of hemoglobin. These days, models of even larger proteins can be built by a single graduate student or postdoctoral researcher in a year.

Perutz and his colleagues, with limited computing power, built their molecules one amino acid residue at a time. They relied on tacit knowledge, their "feel" for molecular structure. Perutz described seeing the molecule emerge as "reaching the top of a mountain after a very hard climb and falling in love at the same time."[18] His comment recalls the sensibility of physicist William Malven who spoke of science as a place where "the mundane and profound go together—like washing dishes and love." In this view of scientific practice, science will always be a human practice, a labor of love that cannot be fully automated. Malven was willing to automate only the most laborious calculations, and then only with the most transparent instruments, as transparent as his Swiss Army knife.

This "human practice" view of science still informs the professional identities of some protein crystallographers. For example, Professor Diane Griffin, head of an East Coast protein crystallography laboratory, uses the phrase *manual thinking* to refer to aspects of protein crystallographic practice that resist full automation, including data gathering, imaging, image analysis, and the calculation of crystallographic maps and models.[19] Griffin belongs to a generation of protein crystallographers who grew up writing their own programs to calculate electron density maps from X-ray data. Her science depends on the accuracy of these programs and she knows how hard it is to get errors out of them. Like Malven, twenty years before her, who spoke of a black box as the most dangerous instrument in a laboratory, Griffin loses confidence when she cannot see inside the programs she uses. At the Spring 2005 MIT workshop, she says:

When I was a graduate student, if you were going to convert some data or something like that, you would write the FORTRAN code to convert the data yourself. That's how you would do it. Now there are these programs. There are these windows and you click. I find with my students all the time, they don't know why something isn't working. I'm like, well, did the data convert properly? Open the file and look at it. It is so black box and it is going from the time when you knew how the data was converted, because you wrote the code to do it yourself, to you don't even open the file to see if it is full of zeros or not. So there is a very big disconnect.

Like Malven, Griffin is particularly skeptical about the use of proprietary software in science; manufacturers have a stake in closing the black box, of keeping code a secret.

Griffin was mentored by a generation of researchers who taught her that scientists should never abdicate authority to instruments they did not fully understand. For them, the advent of opaque software put the scientist in an unacceptable state of ignorance. In a spirit of vigilant skepticism, Griffin educates graduate students, both in her laboratory and across her campus, to exercise critical judgment about computer-generated data.[20]

The field of structural biology includes two distinct groups. Scientists such as Griffin crystallize proteins, conduct X-ray diffraction experiments, and build onscreen molecular models "by hand." A second group of predictive modelers work on complex algorithms to predict protein structure. Those who work with Griffin's methods insist that they need transparent software to achieve their ends. They need to continually adjust and readjust code. The intensity of their involvement keeps the limitations of their representations constantly before them. They are not likely to confuse the model with the molecule. Today, it is these crystallographers who produce the trusted structures against which predictive modelers test their algorithms. The future, however, is uncertain: predictive modelers put their faith in increasingly powerful computers, increasingly powerful algorithms.

Griffin does not trust the claim of predictive modelers, that their software can automatically fit molecular structures to X-ray crystallographic data. She has banned their software from her lab. When she discovered that one of her students had used predictive software to help build part of a model, she made the student repeat the modeling work by hand. And indeed, the computer program had gotten the structure wrong. Griffin's fears had been well founded.[21]

These days, there is intense competition between predictive modelers and crystallographers to be the first to publish protein structures. In competitive science, speed is always of the essence, and this pushes the field toward greater use of automatic techniques. But for Griffin, the automation of model building is a kind of futile cheating: it provides a shortcut that might get you to the wrong place. And even if it brings you to your destination, automation may shortchange you. It certainly shortchanges students because it does not teach them how to use simulation with vigilance. It deprives them of some fundamental experiences they need to develop a tacit knowledge of molecular configurations. Griffin thinks that crystallographers learn to "think intelligently about structure" by slowly building onscreen models. To do this, protein modelers must learn to work intimately with the computer, building a new hybrid instrument, a "human-computer lens."[22]

MATERIALITY IN IMMATERIALITY

In today's biology, computer simulations are ever more manipulable, ever more easily experimented on. They offer an interactivity that makes screen objects seem "material" to the point that contact with them feels like engagement with something quite real. Traditionally, scientists rely on "witnessing" and "participation" to make claims for the legitimacy of scientific knowledge.[23] Familiarity with the behavior of virtual objects can grow into something akin to trusting them, a new kind of witnessing. It is a different sort of trust than Diane Griffin requires, but it can come to feel sufficient.

Griffin began her graduate training in the late 1980s when trust in one's computational tools was associated with familiarity with their underlying code. Younger scientists are increasingly comfortable with black-boxed simulations. They grew up with personal computers that did not come with programming languages. They grew up on computer games that offered interactivity without transparency. Unlike a previous generation, they did not program their own games. When these younger scientists work with screen molecules, they are more likely than their elders to give themselves over to feeling in the grip of a new materiality.

In this they share an aesthetic with the architects who "fly through" virtual buildings. In architecture, models of buildings are rotated on the screen. In biology, molecular models are rotated on their axes. Through these actions, molecules are kept in motion so that the hidden parts of the structure can be brought into view. In both cases, the experience of depth is suggested by performances that engage the body.[24]

Although generational markers are important, in design today, attitudes toward simulation do not neatly sort by generation. The same is true for science. Youth does not automatically confer uncritical comfort with what simulation offers. And age does not automatically lead to resistance to simulation. Some older scientists, for example, justify their use of opaque software by pointing to the infinite regress of computer representations. After all, they argue, it doesn't really mean much to know how your simulation is programmed if all you are looking at is a high-level computer language. The "real guts" of the program is in assembly language and in all that lies beneath that, and no one wants to go to that level with today's complex machines. In the 1980s, Professor Barry Niloff insisted that his students learn the physics of display technology; today, such scruples seem of a different era, practical impossibilities that lead scientists, young and old, to accept opacity. These days, the problem for the working scientist boils down to a question: What level and language will provide enough understanding for me to compare the simulation before me with what I know of nature?

Some younger scientists who are not altogether content with their opaque simulations feel they have no way to act on their unease. One, a physicist at a national research laboratory, admits that when he works with new, elaborate 3D simulation, he misses the algorithmic understanding he enjoyed with earlier models. An older colleague encourages him to play with immersive virtual realities in the spirit of a tinkerer. Time and interaction will do their work: "Give yourself a few years to try it out and fiddle with it awhile," he says. "You will probably find something you can do that you couldn't do the other way." He is convinced that at some point his younger colleague will feel at one with the technology; he will come to "see in simulation," despite its opacity.[25]

Gordon Research Conferences provide an international forum for the presentation and discussion of frontier research in the biological, chemical, and physical sciences. At a Gordon conference in 1965, the structural biologist Robert Langridge presented an interactive computer graphics workstation for visualizing and manipulating molecular models to an unenthusiastic audience of his peers. Langridge recalled that the objections had to do with people not having their "hands on something, something physical so that [they] could understand it." He was not discouraged. In contemporary terms, molecular biology did not yet have the right "interaction metaphor." He said: "Standing up at a conference and showing 16mm movies, in the early days, was really not a good substitute for sitting in front of the computer and actually using it." Even though the early simulations were slow, they made it clear that screen molecules could be compelling: "When you first got your hands on that crystal ball at Project MAC and moved the thing around in three dimensions it was thrilling. There was no question."[26]

As the virtual became increasingly manipulable, as screen movements seemed to happen in real time, protein crystallographers became willing to make the transition from physical to virtual models. With the new technology, one had the sense of dealing directly with the molecule, a feeling that did not depend on the model's appearance, but on the smoothness of the user's interaction with the screen representation.[27] In 1977, a molecular graphics system called GRIP (Graphics Interaction with Proteins) reached a turning point in fluidity of use. GRIP gave its users more than an illusion of smooth connection between modeler and molecule; users experienced the system as a prosthetic extension of themselves into what felt like a tangible world of screen molecules.[28] It is an effect that is familiar to all who play computer games.

In a lecture on the role that simulation plays in protein crystallography, Griffin describes the physicality of today's modeling systems. The best-designed modeling systems try to give protein researchers the tactility and immediacy they came to expect in molecular modeling work.[29] A user sits in front of a screen, often wearing stereoglasses to enhance three-dimensional effects: "You are physically dragging pieces of protein structure, amino acids, and sticking it in [the screen molecule]. You drag it in and you stick it there. And then with your dials or your mouse, you are adjusting it, moving the pieces to get it to fit. So you are physically building with the stereoglasses and the mouse."[30] Protein crystallographers report that they feel the model in their bodies and that their bodies mirror the models they manipulate onscreen.[31]

It is not surprising that, in this relationship with the computer, individual scientists express individuality and differences in style. Some scientists want to use the most up-to-date tools, but many enjoy the comforts of the most familiar well-worn virtual tools.[32] Griffin says that, in her laboratory, researchers tend to use the programs they built or the programs they learned on. Griffin herself uses a program she wrote herself: "Because I'm so familiar with it, I can just do things automatically, which with another program I would have to sit there and think. . . . I've connected with the software in a way that I don't have to think about the direction. . . . I just kind of know how to move the mouse to do what I want to do without thinking."

Across the professions, software has become increasingly uniform and black boxed, even as there is demand for nonstandardized tools that can accommodate users with different intellectual styles. Since today's users cannot change fundamental things about their programs, many return to what in the 1980s was called "customization," small changes that make people feel more at home.

These days, life scientists do not talk much about moving from physical models to computer screens—that ship has sailed. Now they talk about the stress of moving among different virtual environments. Griffin knows that "forcing people to use a uniform program" doesn't work. Researchers are searching for a subtle connection, to make of the software what the psychoanalyst D. W. Winnicott would call a "transitional object," an object that is experienced as separate but also as part of the self.[33] Griffin describes the delicate dance of scientist and choice of simulation in terms that evoke Winnicott. She says that biologists need to work with individualized software, "because some people find certain kinds of manipulations easier. Or just the way a program is organized just works with their brain better." The well-worn or best-loved virtual starts to take on some of the qualities of the real. It feels familiar, comfortable; it is able to assuage anxieties about being cut off from nature.

ENGINEERING THE LIFE SCIENCES

In the life sciences, classically trained engineers, experts in simulation, have created a place for themselves alongside biologists. The engineers, with their expertise in structures and mechanics, bring a distinct way of looking at nature, one that hopes to quantify. Their aspiration is to someday design and build their own molecules and synthetic cells. Life scientists have long used metaphors drawn from engineering and design, as for example, when they referred to proteins as "molecular machines."[34] But current simulations bring them closer to algorithmic descriptions of life.

Today's engineer/life scientists are frustrated that biology is information-rich but data-poor. Its experiments are highly specific and this makes it hard to share data, to build a quantified "meta-model." The engineers push for more shared conventions; one engineer in the MIT biology department speaks wistfully of a "service manual for representing information." His goal is a kit of parts that would let him design new biological systems in simulation. Thus formalized, one could "mine" biological models for data, all of this a dream that requires engineers to "organize the parts, the rates, the components." These aspirations recall those of architects who dream of putting the building inside a machine, of becoming its "geometer."[35]

If biology wants to take on these kinds of goals, it will need engineering-style standards for how it codes and communicates information. Traditional biologists fear that such standards will change what they look for when they look at life—they worry that the biologist's vision will be shaped by the standards that simulations demand. At the Spring 2005 workshop Griffin describes feeling a disconnect between herself and one of her research partners, a biologist with an engineering and computer science background. In a common project, the engineer and his students were to create a simulation of a protein molecule at its lowest possible energy level. Their program produced a result, and Griffin describes them as "proud of themselves," for "they had gotten this fabulous low-energy structure." But when Griffin checked their result against her understanding of proteins, she realized that her colleagues were suggesting a molecule that could not exist.

I tried to explain to them that proteins don't look like that. What they had created did not exist except in sort of a proteosome that was degrading it but this was not a structure. I got them books and [showed them] what an alpha helix was and all this stuff and I finally gave up. There was no ability for us to communicate because they were bound. Their program told them that this was the lowest energy and they were not going to listen to me.

Her engineer colleagues see a result; Griffin tries to interpret its relevance. In her view, the engineer/modelers did not have a sufficiently rich appreciation for biological systems; they did not understand the system's constraints. Their result was beautiful, but its referent was the simulation on its own terms. For the engineer/modelers, the logic of the simulation had overtaken the logic of nature. What Griffin is calling for here is an acknowledgment by her engineer colleagues that simulation's results need to be discussed in light of her understanding of how molecules can look. Her contention was that molecules could not look as they had been represented in the simulation. But there was a barrier to communication, "they were bound." In telling this story, Griffin describes an "enormous divide" between herself and her colleagues from engineering and computer science. They could not put themselves in a position to "check the computer." At the limit, from her point of view, they lost interest in the molecule when it challenged their simulation.

In the early days of Athena, when engineers spoke about a "sacred space" that should be protected from simulation, they identified the analysis of structure. It is telling that in Griffin's efforts to communicate with her engineer colleagues, structure was the first thing to which she turned. ("I got them books and [showed them] what an alpha helix was.") But the engineers she was dealing with were well past trying to reserve the understanding of structure for the tacit knowledge of the experienced scientist.

Some would celebrate the exhilarations of "remediation," translating the gestures of the physical into the virtual, as though what is remediated is illuminated. In the 1980s, simulations let you manipulate what was on the screen; more recently, simulations encourage you to inhabit worlds, or as the architect Donna Gordon put it, "fall into them." These systems are powerful but require a new discipline. We have seen architects and contractors who do not check computer printouts against the reality of their sites and scientists who have a hard time looking up from their screens.

From the earliest days, simulation seduced. In the 1980s, Ted Randall worried that his MIT design students were composing for the screen—its constraints now dictated vision: "I couldn't work with that many contours," said the student who couldn't make screen reality match what was on the ground. If this meant that, in the simulation, twenty-five feet of a site were unaccounted for, so be it. Twenty years later, Griffin's engineer colleagues would not even entertain the notion that their program could be wrong. With the computer on hand to deliver the real, simulation can seem world enough.

NEW WAYS OF KNOWING/NEW WAYS OF FORGETTING

Twenty years ago, designers and scientists talked about simulations as though they faced a choice about using them. These days there is no pretense of choice. Theories are tested in simulation; the design of research laboratories takes shape around simulation and visualization technologies. This is true of all fields, but the case of nuclear weapons design is dramatic because here scientists are actually prohibited from testing weapons in the physical real.

In 1992, the United States instituted a ban on nuclear testing.[1] In the years before the ban, frequent physical tests, first above ground and then underground at the Nevada Nuclear Test Site, provided weapons designers with a place to do basic research. Through tests they developed their scientific intuitions even as they reassured themselves that their weapons worked.[2] More than this, the tests compelled a respect for the awesome power of nuclear detonations. Many testified to the transformative power of such witnessing.[3]

In the years after the 1992 ban, newcomers to the field of nuclear weapons design would see explosions only on computer screens and in virtual reality chambers.[4] At Lawrence Livermore and Los Alamos National Laboratories, some of the most powerful computer systems in the world are used to simulate nuclear explosions. Until recently, these simulations took place in two dimensions; now, simulations are moving into three dimensions.[5] In a virtual reality chamber known as a CAVE, one stands "inside" a nuclear explosion wearing 3D goggles, in order to observe it, one is tempted to say, "peacefully."[6] My story of simulation began with the Athena project centered in a garden, a glass atrium with a ficus tree; it ends in a CAVE, a self contained virtual reality. The CAVE simulation is there to "demo" an explosion; those who work there become accustomed to experiencing in the virtual what could never be survived in the real.

When nuclear testing moved underground, it became easier for weapons designers to distance themselves from the potential consequences of their art. Hidden, the bomb became more abstract. But even underground testing left craters and seismic convulsions. It scarred the landscape. Now, with explosions taking place on hard drives and in virtual reality chambers, how much harder will it be for weapons scientists to confront the destructive power of their work and its ethical implications?[7] One weapons designer at Livermore laments that he has only once experienced "physical verification" after a nuclear test. He had "paced off the crater" produced by the blast. It changed him forever. His younger colleagues will not have that.[8]

This senior scientist is concerned about the moral effects of moving nuclear weapons research to virtual space, but he and his colleagues are also troubled about the effects of virtuality on their science itself. They argue that "physical intuition is a skill you want to keep" and worry that the enthusiastic reactions of young designers to new, flashy virtual reality demonstrations are naïve. One says: "The young designers look at anything new and they say, 'This is so much better than what we had before. We can throw out everything we did before!'" Senior scientists at the national laboratories describe young designers immersed in simulation as "drunk drivers." Within simulation, the happily inebriated show less judgment but think they are doing fine. Dr. Adam Luft, a senior weapons designer at Los Alamos, shows sympathy for the young designers: the new rules compel them to fly blindly. They cannot test their weapons because they must work in the virtual and they are given computer systems whose underlying programs are hard to access. Luft himself feels confident only if he is able to access underlying code. He is frustrated by the increasingly opaque simulations of his work environment. When something goes wrong in a simulation, he wants to "dig in" and test aspects of the system against others. Only a transparent system "lets [me] wander around the guts of [a] simulation." He is wary of making any change to a weapon without personally writing its code. Luft worries that when scientists no longer understand the inner workings of their tools, they have lost the basis for trust in their scientific findings, a concern that mirrors those of MIT designers and scientists of twenty years before.[9]

Across professions, successful simulation gives the sense that digital objects are ready-to-hand. Some users find these interfaces satisfying. Others, like Luft, focused on transparency, are not so happy. They look askance at younger designers who are not concerned about whether they wrote or have even seen underlying code. One of Luft's colleagues at Los Alamos describes his "fear" of young designers: "[They are] good at using these codes, but they know the guts a lot less than they should. The older generation . . . all did write a code from scratch. The younger generation didn't write their code. They grabbed it from somebody else and they made some modifications, but they didn't understand every piece of the code." He speaks with respect of "legacy codes," the old programs on which the new programs are built. "You can't throw away things too early," he says. "There is something you can get from [the legacy codes] that will help you understand the new codes."

At Livermore, a legendary senior weapons designer is about to retire. At the Spring 2005 MIT workshop, his colleagues discuss this retirement and refer to it as "a blow." They are anxious about more than the loss of one man's ability to make individual scientific contributions. He has irreplaceable knowledge about the programming that supports current practice.[10] His colleagues fret: "He has such a great memory that he hasn't written down lots of important stuff. How will people know it?"

The response to this scientist's imminent retirement is a movement to videotape him and all the other scientists who are about to leave service. This will be no ordinary oral history. It is infused with anxiety. Those who know only the top layer of programs feel powerful because they can do amazing things. But they are dependent on those who can go deeper. So those who feel most powerful also feel most vulnerable.

Nuclear weapons design is divided by dramatic generational markers: some designers grew up with routine underground testing, some glimpsed it, some have only experienced virtual explosions. Some designers were trained to program their own simulations, some simply "grab code" from other people and are unfazed by the opaque. Yet when Luft sums up attitudes toward simulation in his field, he makes it clear that the wide range of opinion does not reduce to simple generational criteria. The cultures of weapons laboratories are also in play. For example, at Livermore, older weapons scientists who were very hostile to simulation became far more positive when the laboratory adopted a new metaphor for weapons design. Livermore began to liken weapons design to bridge building. According to this way of thinking, engineers do not need to "test" a bridge before building it: one is confident in its design algorithms and how they can be represented in the virtual.[11]

At Livermore, the change of metaphor made simulation seem a reasonable venue for weapons testing. And at Los Alamos, there are younger scientists who find themselves eloquent critics of immersive virtual reality displays. One says: "I was so attuned to making plots on my computer screen. I was surprised at how little new I learned from [the RAVE]." (The RAVE is the nickname for Los Alamos's virtual CAVE technology.) This designer complains about not being able to work analytically in the RAVE; others say that it gives them a feeling of disorientation that they cannot shake. In the RAVE, scientists work in a closed world with rigorous internal consistency, where it is not always easy to determine what is most relevant to the real.[12] For some younger scientists, even those who grew up in the world of immersive video games, the RAVE seems too much its own reality.

Across fields, scientists, engineers, and designers describe the gains that simulation has offered—from buildings that would never have been dared to drugs that would never have been developed. And they also describe the anxiety of reality blur, that "breaking point" where the observer loses a sense of moorings, bereft of real world referents and precedents.[13] And the very complexity of simulations can make it nearly impossible to test their veracity: "You just can't check every differential equation," says Luft. He pauses, and says again, "You just can't, there are just too many." In nuclear weapons design you can make sure that you have solved equations correctly and that your system has internal consistency. In other words, you can "verify." But he adds, "validation is the hard part. That is, are you solving the *right* equations?" In the end, says Luft, "Proof is not an option."

PRETTY PICTURES

At the Spring 2005 MIT workshop, astrophysicist Peter Charles tells a story of a beautiful image he produced in simulation. It was beautiful, but it did not correspond to anything in the physical real. Charles was working on a scientific problem, but then he "made a mistake," and a compelling image emerged when he plotted his mistake. The image, says Charles, "was very pretty." He plotted an error and created an image that "looked cool but was wrong." There was no question of using the image in a scientific publication—it did not refer to anything real—but it was so visually elegant that Charles could not resist putting it on his personal Web site. There, it attracted the attention of a television network and a scientific funding agency—one used it as a logo, the other for publicity posters. The beautiful but meaningless child of simulation was now traveling in the world as an icon of science. Charles had published it as something beautiful; it was read as something real, something scientific. Charles says, "I see these posters and I cringe."

Charles is a distinguished scientist. What encouraged him to post an image that only had meaning within the world of the computer? At the MIT workshop, Charles does not excuse what he did, but tries to explain how it happened. He explains how long it took him to create the image ("I only put it on my Web site because we had spent all this time running it"). And he talks about the image's beauty ("I made a great picture and somehow that sells"). As a scientist, he knew he had to "let it go." But as a curator of images, he did not want to let it go; he may cringe when he sees it displayed as science, but there is the awkward pride of having made something so fine.

Twenty years before, MIT's Ted Randall argued that one feels at a distance from computer printouts. A designer or scientist will not feel the same kind of connection to something that is not inscribed with his or her own "marks." In a certain sense, Charles did not fully identify with the beautiful image as *his* "mark," for it was the computer that had made it so seductive.

When Charles tells his story at the workshop, his colleagues understand how displaying something beautiful on a personal Web site could feel like a statement of artistic rather than scientific appreciation. The image had not crossed any boundaries. Born in cyberspace, it had stayed in cyberspace. But they all knew how quickly simulations travel. They take on a life of their own. Simulation's pretty pictures are routinely used to persuade nonexpert audiences.[14]

At the workshop, Charles admits that his experience with his beautiful image had left him discouraged. "You can sell anything if you dress it up correctly. . . . You can give a result which is complete 'garbage' but taken out of context, reviewers can't tell the difference." He asks the group if his story about "pretty pictures" had relevance to other fields.[15] Around the table, his question is answered with a resounding yes. Tom Kinney, a professor of aeronautics, points out that sometimes pretty pictures can be deployed to disastrous effect. Airplane pilots get "fixated on their displays and shooting down the wrong things because the displays are so compelling." The architects respond to Charles's question—"Could a good rendering sell a bad design?"—with the story of the hotel with the faux limestone facade, a case where something designed in simulation became a problem when it became an emergency.

In the early days of Athena, architects used colored pencils to prettify computer printouts. After twenty years of technological refinement, things have tipped in the other direction. Now technology persuades with elegant computer-generated images. Across disciplines, researchers resent that they are encouraged to spend energy producing such images. Echoing the concerns of architects who create beautiful computer drawings of buildings not yet designed, Diane Griffin complains that in protein crystallography beautiful images mislead because they imply a finished result even when research is at an early stage. Ribbon drawings of the backbone of protein structures used to take a long time to produce; while they were being developed, their rough look reflected that the scientist was not "quite sure of everything yet." Only a fully determined structure would get a "fancy picture." No one would invest the time to make a beautiful drawing of unproved work. "Now," says Griffin,

you can make that fancy picture in two seconds. The program spits out pretty pictures and when you show that picture, people go, "Oh it's all done!" And you can stand up there and say, "These are sort of the distances but don't believe them. There are big error bars! It's not finished yet! This is a rough idea!" And they'll just hold on to it and go: "This is done because look how pretty it is." So we now on purpose make ugly figures to show it's not really done yet because they don't listen to you when they see it with their eyes [laughter in the background]. You have to show them something ugly if you don't want them to set on it and have it be the truth forever.[16]

On hearing how Griffin intentionally degrades her images to convey lack of certainty, Luft points out that some scientists in his world do the opposite. They use simulation to dress up the not-yet proven so that it looks true.

They don't put the caveats in so you can't call them out, but they make it so pretty that everybody believes it. . . . Let's be honest, sexy images sell. A good portion of my work I based on being able to present a sexy picture. I was talking to a research sponsor who was at a research organization that shall remain unnamed. They told me that the next visualization that I gave them had to sparkle.

Luft would not try to fool peers with "sexy images." But his funders want and, in fact, specifically ask him to produce something sparkly.

Scientists and designers at the Spring 2005 MIT workshop are conflicted in their relationships to the products of simulation. Charles feels disconnected enough from his beautiful image to publish something he knows to be meaningless. Griffin wants to disconnect from her pretty pictures because she finds them too convincing. She adds ugliness to the products of her simulations to signal that they are not-yet-proven. Luft disconnects when he adds sparkle to sell. But even when scientists feel alienated from the demands of their audiences, pleasing them is gratifying. Successful simulations flatter even those who are most critical of them.

FROM THE GARDEN TO THE CAVE

We began with a question inspired by Louis I. Kahn: "What does simulation want?" We have seen what simulation seems to want—through our immersion, to propose itself as proxy for the real.

The architecture faculty who designed Project Athena's Garden dreamed of transparent understanding of design process; today scientists are reconciled to opacity and seeing only a CAVE's shadows. Over the past twenty years, simulation has introduced its dazzling environments and we have been witness to our own seduction. A mechanical engineer instructs his students: "Don't be fooled by the graphics."[17] Luft says that beautiful codes promote the "illusion of doing really great science." Kinney teaches "human supervisory control" to inoculate students against the flashy colors and confusing styles of air traffic control displays. When simulation pretends to the real, buildings look finished before they have been fully designed and scientists find no fault in "impossible" molecules that could only exist on a screen. Computer precision is wrongly taken for perfection. The fantasy, visceral in nature, is that computers serve as a guarantor, a "correction machine." Kinney puts it this way: "As technology becomes more and more sexy, the problem is that we get lured into it, the seduction, and we actually come up with what we think are good displays but actually they're bad."[18]

But scientists such as Luft show us another side to what simulation wants. Perhaps we could say, with no irony, it is what simulation really wants—not to replace the real but to reveal it. Luft describes the paradox of simulation used in this way: "I know the simulation isn't right, but because I have the simulation of something tested and the results, I can make adjustments and prophesies about how it was wrong."

I ask Luft how he tells his laboratory director that "simulations are wrong." How does Luft confront him with this subversive reality? Luft responds: "The polite way to articulate that is that a *single* simulation that is not validated by applicable data cannot be trusted." What speaks most loudly in his answer is what Luft does not say. He knows that the problem is not with a single simulation, and he believes that his laboratory director understands this as well. Together they work with simulation and devise fictions around its use. As Luft says, "Simulations are never right. They're all wrong. Forget it. That's it. They're wrong. Guaranteed. There is more entropy in the real world then there is in your computer. That's just the way it is." Nevertheless, every year, he says, "you can use all that data" from simulation and put it in an "annual assessment document" and "every year the lab directors tell the president that everything is cool. . . . That's what we're doing . . . and the punch line is that all simulations are wrong, thus far."

When Luft says that simulations are wrong, he means that they are incomplete. When he places simulation alongside the real, it is to throw the real into sharper relief; simulation's errors sharpen his view of where the real resides. But, like the inhabitants of Plato's cave, Luft, in his own CAVE, knows reality through the shadows it casts. He describes how he makes those shadows work for him. Luft does not see simulation as a way to see what is "true," but to engage in a dialogue with code. "One of the major skills [in simulation] is being able to identify additional simulations you can run which will determine whether the code [you are working with] is behaving reasonably . . . or whether there is some sign of a bug or a mistake." Similarly, MIT biologist Dean Whitman insists that you need a simulation to produce error so that you can test it against reality, to figure out how it is wrong. If you get the simulation right, you will never understand how it is right. You need it to be wrong and you need to figure out how it is wrong.[19]

Whitman, like Luft, articulates a discipline of extracting information from inaccurate models. Both approach simulation as a trusted error-making machine. An inaccurate model generates an interesting hypothesis, which can then be tested. In the Fall 2003 MIT workshop Whitman sums it up by saying that his research simulations are not represented by pretty pictures but by "uglier and more complicated ones. They have more spaghetti hanging off but they are really useful for research." When Whitman and his colleagues confront these ugly pictures, they expect to take simulation error and do something constructive with it. From Garden to CAVE, the notion of a "critical stance" toward simulation has been transformed. These days, for simulation's most sophisticated users, a critical stance is no longer about vigilance to protect simulation from error. It is about living with shadows that bring us closer to the forms beyond them.

Whitman sums up this point when he talks about the necessity of being very clear about what simulation cannot do. It cannot keep you open minded. The scientist must always ask, "To what extent does a model limit us to iterations rather than opening our minds to new questions?"[20] As a scientist, one must attend to what lies beyond any model:

When you have a hammer, everything looks like a nail. . . . So when you have either a model or a certain capability, and you come to work every day, or you start to write a proposal . . . you say to yourself, "Gee, what kind of crazy blue sky idea can I come up with today?" or "Gee, I think there are some more nails out there and maybe we should start hammering." That's the way I see it. It's a trade-off in the lab how much of one or the other we do.

For Whitman, the hard work begins with resistance to pretty pictures. In the Fall 2003 MIT workshop, he is asked to describe the emotional power of molecular visualizations. Whitman insists that what is most important is to be inoculated against their buzz: "When I started . . . people would show pictures of biological models and say, 'Now we understand.' And I would say, 'No, we don't understand. We have pictures and have the beginnings of something you can use to understand.'" Whitman works in an informed partnership with simulation. It generates alternate realities and enables him to do experiments that would otherwise be impossible. But the limitations of these experiments humble him. Whitman makes progress by chastening simulation, by increasing his understanding of what it cannot tell, and in the end, deferring to human judgment: "I really need a human being to understand why the model says what it is saying and to evaluate that."

In response to Whitman, Professor Roberta Drew, an organic chemist, presents her view of chastened simulation. Drew uses complex probes to determine the forces and energetic fields within molecules. She appreciates the place to which simulation has taken her discipline: "This has given to the microscopist and chemist— I don't want to say a 'godly sense'—but a sense that you can now go in and one-by-one, engineer your molecules or touch the molecules." But she acknowledges that her deepest understanding does not come from her models: "How many times," she asks, "have you heard the story of someone musing about a truly inspirational vision [coming] to them while they were staring at clouds?" She describes the moment of understanding where "totally out of context" one has a thought, "not consistent with going over the model again and again," and indeed, "a bit adversarial to the iterative model, something that comes out of seemingly nowhere." At that moment, we are left godlike, childlike. Understanding comes out of simulation, out of discontents, and out of nowhere.

NOTES

PREFACE AND ACKNOWLEDGMENTS

1. Susan Sontag, *On Photography* (New York: Picador, 1977), 3.

2. Ibid.

3. The work on the Athena Project, funded by the MIT Office of the Provost, extended from 1983 to 1987. The project's work is summarized in Sherry Turkle, Donald Schön, Brenda Nielsen, M. Stella Orsini, and Wim Overmeer, *Project Athena at MIT* (Cambridge, Mass.: Massachusetts Institute of Technology, 1988). The National Science Foundation study extended from 2002 to 2005. The project's work is summarized in Sherry Turkle, Joseph Dumit, David Mindell, Hugh Gusterson, Susan Silbey, Yanni A. Loukissas, and Natasha Myers, "Information Technologies and Professional Identity: A Comparative Study of the Effects of Virtuality," in *A Report to the National Science Foundation on Grant No. 0220347* (Cambridge, Mass.: Massachusetts Institute of Technology, 2005). Any opinions, findings, and conclusions or recommendations expressed in these reports are those

of the authors and do not necessarily reflect the views of MIT as an institution or of the National Science Foundation. The same disclaimer applies to my writing in this volume.

4. This collection reports on many distinct ethnographies that proceeded with different understandings about the anonymity of participants. It is the convention of this volume to provide anonymity to all informants.

SHERRY TURKLE, SIMULATION AND ITS DISCONTENTS

What Does Simulation Want?

1. All participants in the several studies that led to "Simulation and Its Discontents" are granted anonymity, usually, as here, by simply identifying them as professor or student, or as a practicing scientist, engineer, or designer. When particular individuals take ongoing roles in my narrative, I provide them with pseudonyms for clarity.

2. In the preface and acknowledgments I describe these two studies. The first was focused on the MIT experience; the second reached out to the professional community as well as the academy and included two MIT workshops on simulation and visualization, one in Fall 2003 and a second in Spring 2005.

3. On controversy as a window in science studies, see Dorothy Nelkin, *Controversy: Politics of Technical Decisions* (New York: Sage, 1984).

4. This is a paraphrase. The exact citation is: "When you want to give something presence, you have to consult nature and that is where design comes in. If you think of brick, for instance, you say to brick, 'What do you want, brick?' And brick says to you, 'I like an arch.' And if you say to brick, 'Look, arches are expensive and I can use a concrete lintel over you, what do you think of that, brick?' And brick says to you, 'I'd like an arch.'" See Nathaniel Kahn, *My Architect: A Son's Journey* (New Yorker Films, 2003).

5. Several of the contributors to the 2003–2005 study of information technology and the professions have studied these or similar instances. On

architecture, see Yanni A. Loukissas, "Conceptions of Design in a Culture of Simulation" (PhD dissertation, Massachusetts Institute of Technology, 2008); and "The Keepers of the Geometry," this volume. On molecular biology, see Natasha Myers, "Modeling Proteins, Making Scientists: An Ethnography of Pedagogy and Visual Cultures in Contemporary Structural Biology" (PhD dissertation, Massachusetts Institute of Technology, 2007); "Animating Mechanism: Animation and the Propagation of Affect in the Lively Arts of Protein Modeling," *Science Studies* 19, no. 2 (2006): 6–30; "Molecular Embodiments and the Body-Work of Modeling in Protein Crystallography," *Social Studies of Science* 38, no 2 (2008): 163–199; and "Performing the Protein Fold," this volume. On computer imaging in medicine, see Joseph Dumit, "A Digital Image of the Category of the Person: Pet Scanning and Objective Self-Fashioning," in *Cyborgs and Citadels: Anthropological Interventions in Emerging Sciences and Technologies,* ed. Gary Lee Downey and Joseph Dumit (Santa Fe: School of American Research Press, 1997); and Dumit, *Picturing Personhood: Brain Scans and Biomedical Identity, Information series* (Princeton, N.J.: Princeton University Press, 2004). On simulation in aviation, see David Mindell, *Digital Apollo: Human and Machine in Spaceflight* (Cambridge, Mass.: MIT Press, 2008). On nuclear weapons design, see Hugh Gusterson, *Nuclear Rites: A Weapons Laboratory at the End of the Cold War* (Berkeley: University of California Press, 1996); *People of the Bomb: Portraits of America's Nuclear Complex* (Minneapolis: University of Minnesota Press, 2004); and "The Virtual Nuclear Weapons Laboratory in the New World Order," *American Ethnologist* 28, no. 1 (2001): 417–437.

6. For one view of remediation, see Jay David Bolter and Richard Grusin, *Remediation: Understanding New Media* (Cambridge, Mass.: MIT Press, 1999).

7. For simulation as a world unto itself in which other worlds become captured, see Paul N. Edwards, *The Closed World: Computers and the Politics of Discourse in Cold War America* (Cambridge, Mass.: MIT Press, 1996).

The View from the 1980s

1. Winston Churchill. Remarks to the English Architectural Association, 1924, available at <http://www.icf-cebe.com/quotes/quotes.html> (accessed October 20, 2008).

2. See Max Weber, "Science as a Vocation" and "Politics as a Vocation," *From Max Weber: Essays in Sociology*, trans., ed., and with an introduction by H. H. Gerth and C. W. Mills (New York: Oxford University Press, 1946).

3. The programmers of *SimCity* could have made raising taxes lead to more social services and increased social harmony—but they didn't. What is important here is that this young woman did not know how to program in her game environment. She did not know how to change the rules of the game, nor did she consider questioning the rules of the game, or asking if these rules applied beyond the game. She did not know how to measure the program against the history of real cities. In my view, citizenship in the twenty-first century calls for readership skills in the culture of simulation, the digital equivalent to knowing the "Who, What, Where, Why, and How?" of print media. Such readership skills enable people to be critical of simulation, not simply immersed in it. And what is true of citizens can be no less true when citizens act as professionals. For more on this example, see Sherry Turkle, "Virtuality and Its Discontents," *The American Prospect*, no. 24 (Winter 1996): 50–57.

4. Athena was launched in 1983 with a $70 million dollar gift from the IBM and Digital Equipment Corporations for the purpose of using computers in the undergraduate curriculum. Athena's first priority was to create a "coherent network" for educational computing, including hardware, operating systems, and programming languages. The goal was to have a student's one-time investment in learning the system provide access to the full range of educational software developed at MIT. Building on this technically sophisticated network, Athena's architects hoped to integrate "modern computer and computational facilities into all phases of the educational process" in order to "help students learn more creatively and fully in a wide range of disciplines" by developing "new conceptual and intuitive understanding

and [improving and refining] our teaching methods." "Project Athena, Faculty/Student Projects," *MIT Bulletin* (March 1985).

5. In our 1983–1987 Athena study, many MIT faculty believed they could avoid having much to do with computers. This study included interviews with students and faculty as well as observation of courses that had introduced computers into the teaching experience. Four fields were intensively studied. Interviews in civil engineering were conducted by Wim Overmeer and Donald Schön, in chemistry by M. Stella Orsini and Brenda Nielsen; in physics and architecture and planning by Brenda Nielsen and Sherry Turkle.

6. See Seymour Papert, *Mindstorms: Computers, Children, and Powerful Ideas* (New York, Basic Books, 1981), 23.

7. MIT tradition favors teaching programming as an "aside" to courses. At the time of the Athena launch, even the main programming course for computer science majors, 6.001, expected students to teach themselves the LISP programming language and focused on larger issues about the structure of programs. This strategy of expecting students to acquire programming skills on their own worked within the Department of Electrical Engineering and Computer Science. It worked less well in the School of Architecture and Planning. There, anxiety ran high when students realized that they were being left to their own devices on technical matters.

8. On Gehry, see Michael Schrage, "Nice Building, but the Real Innovation Is in the Process," *Fortune,* July 10, 2000. Available at <http://money.cnn .com/magazines/fortune/fortune_archive/2000/07/10/283746/index .htm> (accessed July 21, 2008).

9. For a more complete description of styles of mastery and how computers facilitated diverse styles, see Sherry Turkle, *The Second Self: Computers and the Human Spirit* (Cambridge, Mass.: MIT Press, 2005 [1984]), especially chapter 3; and Sherry Turkle and Seymour Papert, "Epistemological Pluralism: Styles and Voices in the Computer Culture," *Signs: Journal of Women in Culture and Society* 16, no. 1 (1990): 128–157.

10. These days, the sanctity of the lecture hall in the 1980s has a distinct irony. Students now enter lectures and flip open laptops that are directly connected to the Internet.

11. As an electron travels around an atomic nucleus, a magnetic field is set up. Classical physics predicts that, if influenced by another magnetic field, the orientation of the electron will be deflected and there should be a continuum of deflections depending on the strength of the external magnetic field and the magnetic momentum of the atom itself. But quantum physics predicts only two positions; the Stern-Gerlach experiment demonstrates this space quantization.

12. A third experiment in Professor Malven's junior laboratory, the Mossbauer experiment, demonstrated resonance fluorescence, a technique used to infer the mean lifetime of the excited state of nuclei. The technique works in a fairly straightforward way for atomic transitions, but a number of problems arise when you try to use it for nuclear transitions: crystal structures magnify the movement of individual nuclei and widen the energy curve, masking its natural width; when nuclei emit a photon they recoil and this motion shifts the spectrum. Rudolf Mossbauer discovered that by reducing the temperature one could eliminate the factors that skewed the curve. And it was later found that iron (^{57}Fe) could give similarly clear results without lowering temperatures. In the junior lab, the nuclei in a sample of iron were raised to an excited state. As photons were emitted and the energy states dropped, the spectrum was measured by a photometer and the computer was used to record its data and fit it to a curve whose width related to the mean lifetime of the excited state.

13. Physics faculty explained that problems that could be solved by analytical methods (and those methods themselves) had defined "high physics," prestigious physics. The computer made it possible to solve classes of problems that were only accessible through numerical methods, what Professor Niloff referred to as "just plotting stuff."

14. On using emergent phenomena in science pedagogy, see Mitchel Resnick, *Termites, Turtles, and Traffic Jams: Explorations in Massively Parallel Computing* (Cambridge, Mass., MIT Press, 1997).

15. Every laboratory, in some sense, creates an enclosed, artificial environment, a "simulation" necessary to perform experiments. Simulation on the computer takes things further: in their attempts at rigorous internal consistency, simulations become increasingly difficult to calibrate against an external world. See Harry Collins, *Changing Order: Replication and Induction in Scientific Practice,* 2nd ed. (Chicago: University of Chicago Press, 1992 [1985]). Collins and others have used the term "experimental regress" to describe how the calibration of a scientific instrument can become entangled within the highly contrived environment of an experimental framework. In the end, there is no outside source against which to tune the machine. See Andrew Pickering, "The Mangle of Practice: Agency and Emergence in the Sociology of Science," *American Journal of Sociology* 99, no. 3 (1993): 559–589.

16. See Sherry Turkle, *Life on the Screen: Identity in the Age of the Internet* (New York: Simon and Schuster, 1995).

Design and Science at the Millennium

1. In the pages that follow, I report on an NSF-funded study of simulation and visualization conducted from 2003-2005. See Sherry Turkle, Joseph Dumit, David Mindell, Hugh Gusterson, Susan Silbey, Yanni A. Loukissas, and Natasha Myers, "Information Technologies and Professional Identity: A Comparative Study of the Effects of Virtuality," in *A Report to the National Science Foundation on Grant No. 0220347* (Cambridge, Mass.: Massachusetts Institute of Technology, 2005). Two pre-doctoral research assistants were integral to that study, Yanni A. Loukissas, working in architecture, and Natasha Myers, working in structural biology.

Citations from practicing architects and architecture faculty and students in the 2000s draw on interviews conducted by Loukissas and Turkle except where I specify that the citations are from designers participating in one of two interdisciplinary workshops on simulation and visualization held at MIT as part of the NSF study. The first workshop was held in October 2003 (referred to in the text as the Fall 2003 MIT workshop); the second workshop took place in May 2005 (referred to in the text as the Spring 2005 MIT workshop). Some of Loukissas's fieldwork in architectural firms described in the final NSF reported is reported in Loukissas, "Keepers of the Geometry," this volume. When I draw on this work, I cite the final report and the Loukissas paper or his fieldnotes. Loukissas's studies of simulation continued in his dissertation work: "Conceptions of Design in a Culture of Simulation" (PhD dissertation, Massachusetts Institute of Technology, 2008). Loukissas is currently preparing a book-length manuscript based on his dissertation research.

Citations from biologists in the 2000s draw on interviews conducted by Natasha Myers, and reported in the NSF final report, except where it is specified that the citations are from the workshops, as above. Myers's research on the NSF study was the foundation for her dissertation, "Modeling Proteins, Making Scientists: An Ethnography of Pedagogy and Visual Cultures in Contemporary Structural Biology" (PhD dissertation, Massachusetts Institute of Technology, 2007). She is currently preparing a book-length manuscript based on her dissertation research. Myers has published several papers in addition to her dissertation on her structural biology work. When I quote from Myers's interviews that appeared in the NSF report, where possible, I also provide citations from her published work.

2. The extended story of resistance to CATIA is reported in Turkle et al., "Information Technologies and Professional Identity"; and Loukissas, "Keepers of the Geometry," this volume.

3. One design student at the Spring 2005 MIT workshop commented that in large firms there is still room for a "top person" who simply sketches, but most architects starting out and trying to establish their name "can't afford a renderer or a modeler." So today's young architect needs to have simulation literacy "at the cutting edge." This fluency with computation is no longer a "specialization" but a "mode of survival."

4. See Turkle et al., "Information Technologies and Professional Identity" and Loukissas, "Keepers of the Geometry," this volume.

5. On the body in performances of simulation, see Rachel Prentice, "The Anatomy of a Surgical Simulation: The Mutual Articulation of Bodies in and through the Machine," *Social Studies of Science* 35, no. 6 (Dec. 2005): 867–894; Natasha Myers, "Animating Mechanism: Animation and the Propagation of Affect in the Lively Arts of Protein Modeling," *Science Studies* 19, no. 2 (2006): 6–30; "Molecular Embodiments and the Body-Work of Modeling in Protein Crystallography," *Social Studies of Science* 38, no 2 (2008): 163–199; and "Performing the Protein Fold, " this volume.

6. Like high-definition television, what appears on the screen can seem more precise than nature, the "out-realing" of the real. This idea is informed by the notion of a hyperreal, as discussed by Jean Baudrillard. See Jean Baudrillard and Jean Nouvel, *The Singular Objects of Architecture,* trans. Robert Bononno (Minneapolis: University of Minnesota Press, 2002).

7. For a classic discussion of simulations and new classes of objects that simulate objects that never existed, see Jean Baudrillard, *Simulacra and Simulations,* trans. Sheila Faria Glaser (Ann Arbor: University of Michigan Press, 1994).

8. In the history of space flight, moments of reality blur have been dramatic. Historian David Mindell has observed that for months before the flight to the moon, the astronauts lived most of their lives within simulators, flying to the moon under a great variety of conditions, simulated every conceivable kind of failure. Flight simulators blurred the boundary between real and virtual flight, and proved a valuable rehearsal for the human-machine system that would land on the moon. See Mindell, *Digital Apollo: Human and Machine in Spaceflight* (Cambridge, Mass.: MIT Press, 2008), chapter 1. Indeed, at crucial moments during Apollo's actual lunar landing, the experience of simulation and the real seemed to enfold. When astronaut Buzz Aldrin lost contact with ground control in Houston, he needed to make manual adjustments to the automatic system that controlled the direction of the communications antennae. There was static on the line; it

was hard for the ground controllers to hear the astronauts over the noise. As the frustrated controllers struggled to piece together a continuous story from intermittent bursts of data, one said, "This is just like a simulation." Mindell notes that "indeed, the performance had been rehearsed, countless times in countless variations, in computer-controlled virtual simulations on the ground," *Digital Apollo*, 2.

The history of aviation shows how the "real world" comes to be experienced through the lens of simulation. Jean Baudrillard evokes a related notion of reality blur as he describes how simulations may overrun reality, leaving the "desert of the real" in their wake. See *Simulacra and Simulations*.

9. The relationship between architect and craftsperson in simulation culture is complex. In "Keepers of the Geometry," this volume, Loukissas describes a new digital craftsmanship, the foundation of a new relationship between designers and craftspeople.

10. The anthropologist Gary Downey characterizes contemporary designers as having given their "eyes over to the computer as well as [their] fingers." See *The Machine in Me: An Anthropologist Sits among Computer Engineers* (New York: Routledge, 1998).

11. Yanni A. Loukissas, field notes, 2003. Cited in Turkle et al., "Information Technologies and Professional Identity."

12. W. Deane Butcher, quoted in Evelyn Fox Keller, *Making Sense of Life: Explaining Biological Development with Models, Metaphors, and Machines* (Cambridge, Mass.: Harvard University Press, 2002), 16–41.

13. Workshop on Simulation and Visualization in the Professions, MIT, October 2003. In what follows, comments by life scientists are from field notes and interviews by Natasha Myers, research assistant NSF grant no. 0220347, and cited in its final report, unless as here, attributed to presentations and conversations at the MIT workshops on simulation and visualization in the professions in October 2003 and May 2005.

14. Such achievements depend on scientists from many different disciplines coming into the life sciences; it draws on expertise in chemistry,

physics, materials science, computer science, and engineering as means to access new kinds of information about living processes, particularly at the scale of cells and molecules. The life sciences today consist of many overlapping fields of research including biochemistry, molecular biology, cell biology, structural biology and protein folding, genomics and proteomics, and computational and systems biology. A wide array of computer technologies has been applied to the challenges of visualization in the life sciences. These include including microscope images, quantitative assays of gene expression and protein activity, and high-resolution models of molecules from X-ray crystallography, and other modes of structure determination.

15. See Michael M. J. Fischer, *Emergent Forms of Life and the Anthropological Voice* (Durham: Duke University Press, 2003).

16. See Peter Galison on women laboratory technicians as "computors" in physics. Galison, *Image and Logic: A Material Culture of Microphysics* (Chicago: University of Chicago Press, 1997), 199, 375, 718. In crystallography, women on short-term contracts worked through the 1950s, measuring and recording by hand the reflections produced by diffracting X-rays through protein crystals. See Soraya de Chadarevian, *Designs for Life: Molecular Biology after World War II* (Cambridge: Cambridge University Press, 2002), 123. See also Lily E. Kay, *The Molecular Vision of Life: Caltech, the Rockefeller Foundation, and the Rise of the New Biology, Monographs on the History and Philosophy of Biology* (New York: Oxford University Press, 1993). In the early years, protein crystallographers treated the data produced by computers with skepticism. For example, Max Perutz resisted using computers until it became absolutely necessary. See de Chadarevian, *Designs for Life,* 126.

17. See de Chadarevian, *Designs for Life,* and Eric Francoeur and Jerome Segal, "From Model Kits to Interactive Computer Graphics," in *Models: The Third Dimension of Science,* ed. Soraya de Chaderevian and Nick Hoopwood (Stanford: Stanford University Press, 2004).

18. Turkle et al., "Information Technologies and Professional Identity." Cited in Myers, "Molecular Embodiments," 181. Here, Myers works through an extended analysis of "feeling," tacit knowledge, and love and labor in the

craftwork of molecular model building. On these themes, also see Myers, "Modeling Proteins, Making Scientists" and "Animating Mechanism."

19. See Myers, "Modeling Proteins, Making Scientists" for a detailed analysis of Griffin's notion of "manual thinking" and its role in building protein models. Myers also offers a discussion of the pedagogical challenges that protein crystallographers face when trying to cultivate these tacit skills in a new generation of students, and shows how these issues are compounded by the increasing presence of automated techniques in this field.

20. For an extended discussion of the Griffin case and the need for cultivating critical judgment in molecular visualization see Myers, "Molecular Embodiments," and "Modeling Proteins, Making Scientists."

21. For more on this case, see Myers, "Molecular Embodiments" and "Modeling Proteins, Making Scientists," especially 26–32.

22. For a discussion of what it takes to learn how to "think intelligently about structure" see Myers, "Molecular Embodiments." There, Myers also examines how protein crystallographers collaborate with their computers to form a "human-computer lens" to draw protein structures into view. For an early formulation of the "human-computer lens" analogy as it is used among protein crystallographers, see Jenny Glusker and Kenneth Trueblood, *Crystal Structure Analysis, A Primer* (New York: Oxford University Press, 1985), 5.

23. The classic treatment of "witnessing" in science is Steven Shapin and Simon Schaffer, *The Leviathan and the Air-Pump: Hobbes, Boyle, and the Experimental Life* (Princeton: Princeton University Press, 1985).

24. On structural biologists and body performance, see Myers, "Performing the Protein Fold," this volume.

25. Workshop on Simulation and Visualization in the Professions, MIT, May 2005.

26. Robert Langridge, quoted in Eric Francoeur and Jerome Segal, "From Model Kits to Interactive Computer Graphics," 418. Cited in Myers, "Molecular Embodiments," 175.

27. On the physicality of modeling see Francoeur and Segal, ibid.; Myers, ibid. and "Animating Mechanism"; Prentice, "The Anatomy of a Surgical Simulation"; and Yanni A. Loukissas, "Representations of the User in 3D Geometric Modeling," paper presented at the Annual Meeting of the Society for the Social Studies of Science, Pasadena, 2005.

28. Myers, "Molecular Embodiments," 177. See D. Tsernoglou et al., "Molecular Graphics—Application to Structure Determination of a Snake-Venom Neurotoxin," *Science* 197, no. 4311 (1977): 1378–1379.

29. For Natasha Myers's discussion of modelers and this sense of physicality, see "Molecular Embodiments," 186. To examine the intimate association of bodies and tools for learning she draws on the work of Maurice Merleau-Ponty and Michael Polanyi, and points out that Polanyi himself was trained as a crystallographer and develops elements of his thinking about tacit knowledge with reference to crystallographic practice. See Polanyi, *Personal Knowledge: Towards a Post-Critical Philosophy* (Chicago: University of Chicago Press, 1958).

30. Turkle et al., "Information Technologies and Professional Identity." Cited in Myers, "Molecular Embodiments," 178.

31. On cultivating a "feeling for the molecule," see Myers, ibid. On the question of "feeling" as a mode of understanding, see Evelyn Fox Keller, *A Feeling for the Organism: The Life and Work of Barbara McClintock* (San Francisco: W. H. Freeman, 1983); and *Reflections on Gender and Science* (New Haven: Yale University Press, 1999).

32. They feel connected to particular "interaction metaphors," the forms presented by the human-computer interface. A nonscientific example of an "interaction metaphor" is, for example, the desktops and file folders that people use on their personal computers. In "Anatomy of a Surgical Simulation," Prentice uses the term "mutual articulation" to describe the work of simulator researchers in developing effective computer simulations for teaching anatomy and surgery. In this case the designer must articulate the model patient and the body of the surgeon who will use this

interface for training. Only the modeler's body can serve as a template for these entanglements.

33. D. W. Winnicott, *Playing and Reality* (New York: Routledge, 1989 [1971]).

34. On the metaphor of "molecular machines," see Natasha Myers, "Modeling Molecular Machines," in *Nature Cultures: Thinking with Donna Haraway,* ed. Sharon Ghamari-Tabrizi (Cambridge: MIT Press, 2009).

35. On this, see Loukissas, "Keepers of the Geometry," this volume.

New Ways of Knowing/New Ways of Forgetting

1. On the professional culture of American nuclear weapons scientists, see Hugh Gusterson, *Nuclear Rites: A Weapons Laboratory at the End of the Cold War* (Berkeley: University of California Press, 1996); and "The Virtual Nuclear Weapons Laboratory in the New World Order," *American Ethnologist* 28, no. 1 (2001): 417–437.

2. Joseph Masco, "Nuclear Technoaesthetics: Sensory Politics from Trinity to the Virtual Bomb in Los Alamos," *American Ethnologist* 31, no. 3 (2004):1–25. U.S. tests were conducted above ground between 1945–1962 and below ground between 1963–1992.

3. Richard Rhodes, *The Making of the Atom Bomb* (New York: Simon and Schuster, 1988).

4. Nuclear weapons scientists may no longer test weapons and do much of their research in the virtual. At the Spring 2005 MIT workshop on simulation and visualization, the conversation naturally focused on work in simulation. This does not, however mean that all weapons development is done virtually. There are underground experiments where plutonium is exploded without setting off a nuclear explosion, laser facilities at Livermore Laboratories (the NIF) and at Sandia (the Z-Pinch), and a facility at Los Alamos that takes x-ray snapshots of the "pits" inside atomic bombs as they implode (the DARHT)—again without a nuclear explosion. All of these provide indirect ways of getting at what is happening inside a nuclear bomb.

5. The Advanced Strategic Computer Initiative, part of the new Nuclear Stockpile Program, helps the weapons laboratories to build more advanced computer systems that will be able to run codes embodying three-dimensional simulations of nuclear explosions. The old codes, now being updated, were two-dimensional.

6. CAVE is an acronym for Cave Automatic Virtual Environment. The Los Alamos CAVE has been criticized as not much more than a flashy medium for publicity stunts. Its effectiveness as a research tool is an ongoing concern. According to one weapons designer at the MIT workshop in May 2005: "At some point, it [the CAVE] is more for show. I don't know when it will become mainstream. You do it for dignitaries coming in. We did our images for the Prince of York."

7. Political leaders, too, can now witness nuclear events only through simulation. Harold Agnew, director of Los Alamos Laboratory from 1970 to 1979, once said, "I firmly believe that if every five years the world's major political leaders were required to witness the in-air detonation of a multi-megaton warhead, progress on meaningful arms control measures would be speeded up appreciably." Harold Agnew, "Vintage Agnew," *Los Alamos Science* 4, no. 7 (1983): 69–72. These days, dignitaries are brought to watch weapons explode in virtual reality. What this means is that goggles in hand, you come to watch something akin to a multibillion dollar video game whose special claim is that it simulates an existing agent of death.

8. Workshop on Simulation and Visualization in the Professions, MIT, October 2003. All further citations of nuclear weapons designers and researchers are from the May 2005 MIT workshop.

9. Luft wants to have access to programs at the level of code but does not enjoy programming. He sees himself as an exception in his generational cohort and says of his same-age colleagues: "They all love to get in there and go do their programs. I just personally choose not to." Workshop on Simulation and Visualization in the Professions, MIT, May 2005.

10. See Hugh Gusterson, "A Pedagogy of Diminishing Returns: Scientific Involution across Three Generations of Nuclear Weapons Science," in *Pedagogy and the Practice of Science*, ed. David Kaiser (Cambridge, Mass.: MIT Press, 2005).

11. Architects are sensitive about the irony in the use of the bridge metaphor as a source of reassurance about designing in simulation. Architect Sir Norman Foster's Millenium Bridge in London, designed in collaboration with sculptor Sir Anthony Caro and Arup engineers first opened in June 2001. During the first days large numbers of people crossed the bridge producing a pendulum-like sway. The bridge was closed after three days only to be reopened in February 2002 after research and testing resolved the problems. Dampers were installed under the deck and between the deck and the river piers to reduce the sway. See <http://www.bluffton.edu/~sullivanm/england/london/millennium/foster.html> (accessed July 28, 2008).

12. See Paul N. Edwards, *The Closed World: Computers and the Politics of Discourse in Cold War America* (Cambridge, Mass.: MIT Press, 1996).

13. Workshop on Simulation and Visualization in the Professions, MIT, May 2005.

14. On images and how they travel, see Joseph Dumit, "A Digital Image of the Category of the Person: Pet Scanning and Objective Self-Fashioning," in *Cyborgs and Citadels: Anthropological Interventions in Emerging Sciences and Technologies*, ed. Gary Lee Downey and Joseph Dumit (Santa Fe: School of American Research Press, 1997); and Dumit, *Picturing Personhood: Brain Scans and Biomedical Identity* (Princeton, N.J.: Princeton University Press, 2004).

15. Helpful works on simulation and seduction include: Richard Doyle, *Wetwares: Experiments in Postvital Living* (Minneapolis: University of Minnesota Press, 2003); Peter Galison, *Image and Logic: A Material Culture of Microphysics* (Chicago: University of Chicago Press, 1997); Edwards, *The Closed World;* Stefan Helmreich, *Silicon Second Nature: Culturing Artificial Life in a Digital World* (Berkeley, Cal.: University of California Press, 1998);

Evelyn Fox Keller, *Making Sense of Life: Explaining Biological Development with Models, Metaphors, and Machines* (Cambridge, Mass.: Harvard University Press, 2002); Myanna Lahsen, "Seductive Simulations? Uncertainty Distribution around Climate Models," *Social Studies of Science* 35 (2005): 895–922; and Sergio Sismondo, "Models, Simulations, and Their Objects," *Science in Context* 12, no. 2 (1999): 247–260.

16. Workshop on Simulation and Visualization in the Professions, MIT, May 2005.

17. Natasha Myers, field notes, Fall 2003. Cited in Sherry Turkle, Joseph Dumit, David Mindell, Hugh Gusterson, Susan Silbey, Yanni A. Loukissas, and Natasha Myers, "Information Technologies and Professional Identity: A Comparative Study of the Effects of Virtuality," in *A Report to the National Science Foundation on Grant No. 0220347* (Cambridge, Mass.: Massachusetts Institute of Technology, 2005).

18. Workshop on Simulation and Visualization in the Professions, MIT, May 2005.

19. Workshop on Simulation and Visualization in the Professions, MIT, October 2003.

20. Ibid.

SITES OF SIMULATION: CASE STUDIES

Outer Space and Undersea

BECOMING A ROVER

William J. Clancey

It is 3:13 a.m. at Gusev Crater on Mars, and the Mars Exploration Rover (MER) called *Spirit* is powered down for the night. The team of scientists "working Gusev" lives on Mars time, but, with some luck, they are fast sleep in California.[1] The MER is a remotely operated vehicle and it is not the only one exploring Mars at this time. On nearly the opposite side of the planet at Meridiani Planum, another MER called *Opportunity* photographs the Martian surface and undertakes the analysis of minerals. The thirty-seven scientists "working Meridiani" are ensconced in a fifth-floor meeting room at the Jet Propulsion Laboratory in Pasadena. With my background in robotics and artificial intelligence, I am living as if on Mars with this team as I attempt to document its mission. It is a bright February afternoon, but we sit in a room darkened by heavy black shades. We are with our rover. We count time by the number of sols, Mars rotations, since our rover landed. Today is sol 25, M25.[2]

The first meeting of sol 25 includes short lectures by a half-dozen scientists. Speaking with wireless microphones, they display scores of colorful photographs and charts with titles such as "Locations and Things to Do for Mineralogy." The MER scientists sit in clusters, organized into four Science Theme Groups and a Long-Term Planning Group. Later in the sol, the Science Operations Working Group polls the thematic groups for what they want the rover to do, what commands they want to give to the robotic vehicle. These commands will be converted into eight-hour-long software instructions that will program the rover's actions. Before the end of the day, the scientists will settle on one set of instructions, known as a sequence. This sequence will be communicated to an engineering team who will prepare computer code for the rover's next-day exploration of Mars. An end-of-sol meeting will review daily progress, next-day plans, and the group's long-term goals.

Today, on sol 25, a month into the *Opportunity*'s mission, the Chair of the Science Operations Working Group gives a short lecture about the Rock Abrasion Tool, the RAT, which serves as a geologist's hammer, scraping a circle into a hard surface. The speaker warns the group against indiscriminate data gathering with the RAT, letting their curiosity bypass the discipline of scientific hypothesis testing. "As we think about how we are going to approach this outcrop, our thinking and our discussion should be very much based on hypothesis testing. . . . Don't say 'let's RAT here' to reveal this, and 'let's RAT here' to reveal that; let's talk it through in terms of the specific scientific hypotheses that we're trying to test."

Why is it necessary to lecture these scientists—experienced in field exploration, competitively selected to be members of the MER Science Team, many of them having worked on multiple missions before this one—on how to do science? Working with the remote-controlled rover changes the practice of field science in ways that make such lectures necessary, reshaping intellectual practices and professional identities.

AN EXPLORER'S IDENTITY: BECOMING A ROVER

I chose to interview six MER scientists who represent different generations within the space program, spanning from those who have made NASA a lifetime career to the young scientists for whom the MER project is their first mission. Getting a place on a mission team at NASA is a highly competitive enterprise. Regardless of their age, the MER scientists view their work on the project as the culmination of their careers, a scientific identity. Ed Dolan, 67, waited almost 20 years for this opportunity, a period during which he was "quite discouraged of the possibility of ever seeing another Mars mission." For Ned Rainer, 56, the project "represents a consummation of a career at NASA. . . . It provides a degree of closure in many things I felt I was preparing for, on behalf of NASA, some 20 years before they happened." Nan Baxter, 34, describes it as a "dream come true. . . . This has been a calling for me in the same way some people are called to be a preacher or something like that. And I am really, really blessed to be doing this." Oscar Biltmore, 44, also feels comforted and fulfilled to be part of the MER team, the benefit of "just not giving up, having people help you along the way, believing in yourself." Bettye Woodruff, 40, sees being on the MER team as an opportunity that will forever mark her career. "For anybody to receive that phone call. . . . That is definitely a defining moment in somebody's career, somebody's life." She explains how the group consists of the most noteworthy people in the field: "I have to say when I saw the roster on that team I was a little bit scared. . . . It was an honor of course to be with those people."

The three younger scientists on the MER team, Woodruff, Baxter, and Biltmore, grew up in the era of planetary science, always knowing that they wanted to work in the space program; the older three, Dolan, Rainer, and Trainor, came to it after doing other things, geophysics, chemistry, and artificial intelligence. Remarkably, given this diversity, all but two of the scientists have a degree related to geology—and refer to this academic experience as a significant source of common understanding.

Working with the rover changes the scientists' sense of their professional identities. It calls on them to be more collaborative. No scientist gets specific credit for work; individual contributions are subsumed to serving the needs of the rover. Furthermore, disciplinary boundaries are erased in ways that make it hard to get a sense of making traditional "advances in one's field." Dolan describes how the focus on the MER blurs the lines between disciplines:

Most scientists work, you know, in their office, with their heads down, and their communication with other scientists is limited. But when you're on these missions, you've got to work with everybody else. You've got to put your heads together. You've all got to come to agreement . . . and you get to know everybody. Not only the scientists, but the engineers, the whole thing.

Woodruff is conscious of this blending and shifting of roles and ways of working. Like Biltmore, she did her graduate work in the 1980s, focusing on planetary science. But whereas a few years ago she was confirmed in her identity as a planetary geologist, now she thinks of herself as "an explorer." Geology is part of that identity, but it has led her to learn more biology to follow her interest in "figuring out Mars," the MER mission objective. It channels the scientists' energy into what Woodruff identifies as the "explorer's spirit." Baxter expresses the same explorer's identity: "If I had lived 500 years ago I would have been on the ships of Columbus or Magellan. Exploration is in my bones."

Baxter, the youngest member of the MER science team, who did her graduate work in the 1990s, sees herself as an explorer with the "mission bug."

I knew I had the bug, the mission bug, a long time ago. I have not taken the traditional path of getting a tenure position and then moving on into research. . . . I have always wanted to do space science. . . . The reason I wanted to do that is the exploratory aspect of studying other planets, doing space science. I've often said that I do space science because I couldn't join Starfleet. [Two] hundred years ago, I might have been on the Lewis and Clark expedition. I really, really enjoy the discovery and exploratory aspect of this. And to me missions are kind of the pinnacle of that.

Trainor, 54, a computer scientist, one of two engineers on the MER science team, sees his role as making other people's jobs easier and accepts that "sometimes that looks like sweeping the floors." He has redefined his sense of professional self worth in terms of working with the rover. Sometimes the code he writes for MER shows him at his best ("Sometimes it looks like I actually get to do . . . what I was educated to do, what I know how to do"), and even when this is not the case, Trainor is sustained by his identification with mission objectives: "I'm all about exploration of space, and this is doing everything I can to make it work. So that's where I come from."

When they joined the project, the MER scientists could not know in advance how their expertise would fit into the mission of getting the rover to explore the surface of Mars. It was not helpful to think in terms of one's narrow specialty. More useful was a pragmatic dedication to the rover. For Trainor, it was easy to position himself in this way, as serving the rover, "simply trying to make myself useful." Other scientists found the transition more difficult. It entails coming to a new view of one's professional training in which it becomes a backstory, something that brought you a ticket onto the mission but is no longer central. One scientist said that, after a point, the repeated daily scientific work of cataloging data "figuring out . . . well, this rock has more olivine than that one . . . kind of gets boring." What stays compelling is being an explorer, a virtual explorer on Mars. As Biltmore puts it, "We've seen things that no one in human history has ever seen, and I've seen them first! . . . That's what gets me going."

Woodruff recalls wanting to be present for the downloading of data from the rover and coming to work hours before her shift was to begin:

I remember . . . the engineers running down to the science room and saying, 'Hey, Bettye, the data is coming down, do you want to come and see that?' and being there and just looking at something for the first time that nobody on the entire planet has seen before you—this is really exploration. And so, this is why this cannot become at any time routine, this is why I want to stay involved.

Each person I interviewed, from the youngest to the oldest, across all fields of specialization, talked about the importance of finding a niche on the rover team, of finding a way to be useful to the rover's voyage. The skills that they brought to the project did not always serve as their best guide for how they would find this niche. They had to embark on a self-directed, matching exercise, reflexive and ongoing: What can I do here? What are my capabilities? Where can I make a contribution?[3] The scientists developed an understanding of productivity that is not relative to the standard metrics of assessing scientific contributions but to a sense of how well they have served the rover. When the niche is found, each scientist reports a sense of pleasure, even of relief. Biltmore describes the anguish of one scientist who confided in him. "He said, 'Why did I get picked for this team because I'm not going to be of any use?' And then, lo and behold! Three weeks later we land in Meridiani and some of his experience and expertise is immediately valuable." In other words, each participant gets a sense of identity from being part of something historical and important. Each feels compelled to make and remake his or her sense of professional self to feel individually relevant, even as individual contributions are hidden in the culture of how the system operates.

Balancing the desire to "fit in" to this new scientific culture, each person I interviewed pursues a personal scientific project in addition to his or her work on the mission. These personal projects provide a sense of individual accomplishment in a setting that minimizes any individual credit. But also, the scientists are continuing normal professional lives as researchers, in which a career is a kind of enterprise with multiple projects and interests. Facets of personal professional work are interwoven in both small group and large communal projects inside and outside of NASA.

Rainer teaches at a small university and has the satisfaction of bringing her unique expertise to bear. Woodruff and Rainer are dedicated to personal inquiries that integrate the MER data with that of other space missions. Biltmore brings fieldwork from terrestrial geology into his planetary interests. Trainor surprises me when he mentions, "I spend lots of long weekend nights doing . . . software modeling of physics. I'm interested in the process of model discovery and refinement." Trainor, who feels he has given up his professional identity as a computer scientist in the MER mission, uses personal projects as a way to alleviate anxiety about his identity as a researcher. "Because those [personal projects] are part of my life, I'm much less concerned about actually proving in MER that I'm doing things that people would consider research-y."

AN EXPLORER'S SURROGATE

Dolan has participated in every Mars mission, but feels a sense of disorientation now when he works with MER. The rover demands a new perspective: "I spent my whole career looking down from above. And now we're down on the ground. It's a very different experience from what I had before." The technology gives scientists the feeling that they are personally on Mars. Biltmore saw obvious things to do at Meridiani "because everything was so laid out in front of me." At Gusev, wondering where the lakebeds might be, "People started saying, 'It's below our feet.'" Asked how he visualized his work, Biltmore says, "I put myself out there in the scene, the rover, with two boots on the ground, trying to figure out where to go and what to do. . . . It was always the perspective of being on the surface." Yet the new perspective can bring its own disorientations. For Dolan, who was accustomed to the "big picture" he got from an orbital vision, "when we got down on the ground (this is me personally now) I'm a little uncomfortable at the narrow focus of the science." For Dolan, putting the MER output into a larger context takes work.

Working with a MER rover requires tending it day after day for years, planning its actions and interpreting the material it transmits back to earth as it moves a few meters a day or remains nearly stationary for months as it studies a few rocks. The MER scientists become solicitous, devoted to this daily effort, the effort of nurturing the MER. But with the nurturance of inanimate things, comes attachment to them.[4]

Steve Squyres's 2005 study of the rover experience—Squyres is principal investigator of the MER work and a documenter of the experience—has many references to the MER scientists speaking of the rover in the first person and to the physical sensation of "being there" that working with the MER instills. ("As we work our way across the plains [of Mars]. . . . We've arrived at Endurance Crater. . . . Where we're standing now."[5]) I think of this as I speak with Baxter, who sums up the decision that the group faces at the end of each day in Pasadena: "Well, are *we* going to go or are *we* going to stay here?"

The rovers were designed to be the scientists' surrogates on Mars. Referring to the RAT, Squyres says: "Our rover was supposed to be a robot field geologist. When you see field geologists on Earth, they've got their boots, they've got their backpacks, and always, they've got big rock hammers."[6] The microimager on the rover is analogous to the geologist's hand lens; the rover's wheels can be programmed to dig trenches (like scraping your boot in the dirt); a camera (pancam) is mounted on the rover at about the height of a man; a brush on the RAT can sweep away dust. The RAT, brush, and microimager are mounted on an arm that has a robotic elbow. It can reach and bend.

In fantasy, the rover becomes the body of each scientist who works with it. Their sense of connection to the device is visceral.

Beyond this, the rover functions as the body of the science team as a whole. Each day, when the scientists decide on the sequence of actions to translate into computer code, they are instantiating their intentions in the MER.[7] The sequence details what instruments the MER will use, what "observations" it will make, where it will go and for how long. It determines when and where the rover will drive or be idle (to recharge its solar batteries, referred to as a "siesta").

Mission practice requires the scientists to articulate the hypotheses that each observation is designed to test. Articulating these hypotheses forces scientists to negotiate disciplinary interests as they share a limited resource and prioritize research questions when daily confronted with the reality that the mission could end at any time. Yet scientific purposes are sometimes dictated by the technology that makes hypothesis testing possible. The importance of the stream of data that the rover generates is historical. Capturing it can be consuming—to the point that it diverts time and attention away from analysis of the data itself. Being involved in "operations," in daily systematic data collection, is compelling, for one is, after all, exploring Mars. Biltmore's conversation about "operations" is infused with this sense of thrill:

I'm still heavily involved in operations, and I'm geology lead for a week out of every month. . . . Even when you're not the lead. . . . I do it anyway because of the interest and honor of being here, to keep up with it and look at these pictures that are coming down every day. I mean, hell, we're on the surface of Mars!

Baxter says:

You're not only using a very expensive machine, you're using a very limited resource and you're also *doing geology for the ages*. . . . This is the only data that we're going to get from Gusev or Meridiani probably for a very, very long time. . . . We have a big responsibility to make sure that data set is as complete but also as compelling as possible.

Early in the mission, the scientists couldn't be sure if *Spirit* would survive the 1.5 mile journey to the Columbia Hills. As a compromise, the team developed a plan to acquire systematic data (called "ground truth") that would be useful for calibrating orbital observations. This plan addressed the legitimate concerns of the scientists and instrument teams that the "forced march" might prevent *Spirit* from acquiring very much science data before she died. This "moving survey," with key observations repeated every fourth sol, was coined by Rainer "the sol quartet."

Any individual who wants to argue for a special observation is expected to articulate a scientific reason, ideally to present a hypothesis to be tested. In practice, especially when arriving at a new site, such investigations may be simply efforts to characterize materials. They are not hypotheses to be tested, but like the strike of a hammer to open a rock, the gestures of explorers to find out what is there. And thus, in the rover that instantiates the scientists' bodies and intellectual curiosity, one sees the natural extension of their explorers' enthusiasms—enthusiasms that do not wait for well-crafted scientific formulations but enable them to revel in the fact of "being there."

Asked what working with a rover is like, Biltmore first describes his sense of "frustration" with the rover's body. He compares what his body could do on the surface of Mars compared to that of the rover surrogates, "They're so slow and plodding." But on reflection, he says that such comparisons are "unfair," for over time human and robot have merged into one explorer:

These things have been our eyeballs out there and our legs and our arms. . . . These rovers? It's been some kind of weird, man-machine bond [laughs]. It's become an extension of each one of us, our eyes or our hands, our feet. . . . I guess in a way, it's through them that we are tasting, tasting the rock. It's. . . kind of, it has morphed into us, or we've morphed into it.

Biltmore and Woodruff confirm this sense of having become "at one" with the machine. For Biltmore: "You want to just hop over those rocks or hop over that ridge over there and climb it, bang on it, do things." And Woodruff describes a similar experience. When she is asked to describe her relation to the MER, she offers, "What about symbiosis? You have to imagine yourself in the field, what would you do if you were there? . . ." Baxter explains:

Sometimes you'll see people talking about, you know, "get a picture behind us" and you'll see them turn their heads. Again, totally unconsciously, because they're thinking of themselves, if they were in the field, what they would try to be, what they would be attempting to look at.

When Rainer describes the process of "becoming the rover" he does so through a concrete scenario: The scientists will see a geological sample of interest and try to move the rover's body to reach it. They began by using engineering diagrams to mentally inhabit the rover's space, but as Rainer says, "Over time we stopped doing it so much because we began to gain a sense of the [rover's] body. That's definitely projecting yourself into the rover. It's just an amazing capability of the human mind. . . . That you can do that, that you can sort of retool yourself."

These projections, "inhabiting the rover," occur especially as the scientists formalize and visualize plans for the next day in computer simulations.[8] The "science activity planner" program enables commanding the rover by pointing to and labeling images previously

taken by the rover, such that chosen targets are automatically regis-
tered in precise 3D Mars terrain coordinates. Other programs then
convert these instrument and target sequences into specific movements and orientations of the arm and instruments. Through this
tight coupling of image, targeting, and feedback, plans are transpar-
ently enacted into exploration paths, such that the scientists move
over several days from broad panoramas to outcrops to particular
rocks and then to a handful of sharply resolved grains. As the rover
moves forward, the returned images are ever more detailed; the
retargeted cameras enable the scientists to make distinctions that
interest them (Is that rock face over there layered? Oh, indeed, it is
layered—are these wind deposits or layers in a stream bed?) and to
move forward in their understanding of how the rocks and terrain
formed over the millennia.The use of simulation is so prevalent
in the MER mission, it appears, as Rainer says, even in taken-for-
granted informal ways, in their gesture and imagination: "There's
something over there of interest to us. . . . [I] thought about yes-
terday. Can we still see it? What can you see if you looked over
there?" Working also with a duplicate rover in a simulated Mars
terrain at the Jet Propulsion Laboratory, the scientists further simulate how the rover will behave—compensating for the impossibil-
ity of directly touching, seeing, or manipulating the stuff of Mars
itself. In coordinating the rover's work across a variety of physical
and computational models, the projection of the self as being the
rover is an embodied way of synthesizing these disparate sources
of information.[9]

So coupled to the rover's sensing and moving, the scientists
report the rover's motions as that of an entity with intention. "*Spirit*
drove . . ." or "*Opportunity* investigated . . ." As if shackled at the

ankles, the team moves together across the plains of Mars, looking and probing the rocks as one body. (Or course, there is some tension. Sometimes they speak as though they are at one with the rovers' bodies, sometimes as if the relationship were parental, as in this remark by Squyres: "*Spirit* and *Opportunity* have been timid, easily frightened into immobility by small rocks. . . . But the new software should make them smarter and more courageous."[10])

Rainer refers to "retooling" himself, Baxter to "morphing," and Woodruff to a "symbiosis," but they are all referring to the same thing—in a manner of speaking, they have become the rover. Although a craft that lands on a planet's surface, such as the *Viking* in 1976, establishes a presence, a rover is different, says Biltmore, "[It's] like the next best thing to being there." Working from orbit, "You're more removed and remote."

The rover can be in only one place at a time. When the scientists have multiple agendas, difficult decisions must be made. The mind of the rover emerges as the scientists negotiate their differences. Every debate about the rover's actions has to end in consensus. Every debate has to end in a decision about a concrete sequence of computer code that will guide the rover's actions. Each step must be articulated. Biltmore gives this account of the debate about whether to move *Spirit* clockwise or counterclockwise around a small rock formation:

We had one of these telecons, an end-of-sol presentation . . . a big discussion. . . . I was pretty vocal about going counterclockwise around Home Plate. . . . It was probably the most wrenching and detailed discussion in Gusev the whole mission. . . . We ended up going the clockwise direction, primarily, for safety. . . . [One] really can't argue. We're all so concerned about power decreasing, because the winter's coming on. . . . What if you get in there and get these shadows and can't get out?

Woodruff, too, was frustrated with the clockwise/counterclockwise debate: "We are here with the most interesting stuff in Gusev, and there we say, 'Okay, drop the rock hammer and leave!' If we don't get to safe haven, we're going to die."

Woodruff's comment dramatizes the degree of emotional connection between the scientists and what is happening on Mars. The idea that the rover might get stuck is experienced as death. The scientists are so identified with the rover that a well-known Mars astrobiologist said at a conference, "The MER robots are really scientists; they are functioning as scientists."[11] It is telling that the scientists experience their work and themselves as inseparable from a technology that simply does their bidding.

For these scientists, the rover has come to represent the mission as a whole, and indeed, talking about the rover has become shorthand for talking about their scientific work. When they describe their work, they speak of the rover's work. Daily reports rarely report on the scientists' activities, they report on the rover's actions: "*Spirit* continued to make progress on the rover's winter campaign of science observations."[12] Besides the convenience of such verbal shorthand, this way of speaking reflects the scientists' projections of themselves into the machine. When they say, "The rover is exploring," they mean, "We are exploring." Losing one's individual contribution to the collectivity is made bearable because it is shared equally and because each scientist identifies with the rover, a compensatory gratification. Biltmore loves the feeling of "two boots on the ground," imagining what the rover can see and reach; others describe the rover as "send[ing] postcards," "becoming more courageous," "finding evidence," and "exploring." No one scientist is allowed to take individual credit for discoveries on the mission, yet in these phrases, some prideful self-description is projected onto the machine.

INTEGRATION OF STYLES AND THE PUBLIC SCIENTIST

Embodiment in the MER requires new integrations among different styles of thought—among these different scientific sensibilities and between the aesthetics of scientists and engineers. For example, the scientists argue about when to drive and when to stay in a given location, a decision that translates into how many measurements to take at a given location. Dolan sees these debates as usually boiling down to a conflict between the geologists and everyone else, with the geologists usually wanting to "get to some more interesting place" and the chemists characterized as never seeing a piece of "soil or a rock that they didn't want to analyze." Dolan interprets the chemists' passion for analysis as emanating from the new instrumentation available to them: "Because they have these instruments there, they want to use them. And they want to stop at every goddamn rock!"

Biltmore uses the language of his professional identity to express similar frustration when the rover pauses too long over what he considers uninteresting terrain. He says, "I'm not into petrology, I'm a geomorphologist." Biltmore like Dolan, sees the geologists on one side of a great divide: "You could see early on who was a field geologist and who wasn't in the group. Because a field geologist saw the value in driving and looking around, in surveying the land. Whereas a lot of people just wanted to sit there and analyze every little sand grain."[13]

When asked to complete the sentence, "Working with a rover is . . . ?" Rainer talks about "working with a team of people . . . and a lot of them engineers." The MER teams have to integrate scientific sensibilities; they also have to take account of the compulsions of engineers to nourish and protect the rover.[14]

At the Jet Propulsion Laboratory, scientists and engineers on the MER mission often work together, always cooperating and often collaborating on problems. But their physical turfs and roles are as clearly defined as on a ship.[15] The separation of floors and rooms, with different key and badge access, formalizes engineering and science into two interleaved, but parallel activities. The engineers are the car's builders, mechanics, and chauffeurs. The scientists indicate where they want the car to go, where it should stop, and what it should do during each stop. The hierarchy is there; yet at the evening meetings of the Science Operations Working Group, the scientists are reminded of the engineers' concerns and constraints. And the scientists are reminded of their ultimate dependency on the engineers. The working group meets in a room suited to high-level diplomacy. Arranged into a large U-shape, there are tables for representatives of the study groups and tables for the engineers. A mission manager sits at the rear of the room. All tables have red lights on table-mounted microphones. The group's humor centers on control of decision making, about who will operate the equipment. The scientists and engineers joke about their fundamental relationship in which the engineer is the service professional and the scientist is the client. As they jostle over who has custody of the rover, they are trying to determine how much of this guardianship is open for discussion.

Indeed, the assembled group characterizes the effects of their MER work as healing the disciplinary divide between engineer and scientist. They claim that working with the rover has caused scientists and engineers to each think more like the other. Baxter says: "I enjoy those times when I can kind of bridge the gap between science and engineering." Woodruff says that the scientists took the initiative to explain what they were doing to the engineers, "to explain to them what we'd be seeing." As with the negotiation of the disparate identities of the scientists, the scientists and the engineers were willing to pull together in large measure because they felt that they were doing something of historical importance. As Woodruff said: "The magnitude of what's happening up there goes way beyond any personal interest or even group interest. It's just a fantastic mission, and just being part of it, and making it work is the only thing that people were interested in. And if it takes [a] 'take it, give it,' kind of thing, this is exactly what's happening."

Out of the crucible of the MER experience came a public face for the mission, including press conferences, magazine stories, even an IMAX movie. What the rovers do every sol has been documented on the JPL Web site, with representative images and explanations of their significance. On selecting the first photographs of the lander, Squyres said, "We want to make a good first impression on the world."[16] Or, as Dolan put it, "It's incredibly public." He told me that while sitting on a plane, "People would realize: *My God! You're running those rovers!* How incredibly informed everybody was, it was amazing! Just amazing, everybody was involved." Biltmore says, "When I meet somebody and they ask me what I do, I say 'astrogeologist,' and almost immediately I say I'm working on the Mars rovers. 'Cause everybody knows what that is." Woodruff compares their exploration to the voyages that discovered the New World: "five hundred years ago, you would take a boat and discover another place. . . . You would record that, come back and tell your story to a limited number of people. The word was spread at the speed of the horse, or human voice. Today six billion people on January 4, 2004, discovered a new site on another planet. This is human exploration. Not human because humans were there, but because we were ALL there, together, through a robot!"

Biltmore, Trainor, and Woodruff told long, detailed stories about the night *Spirit* landed on Mars. Woodruff describes *Spirit's* landing, "At that point . . . we jumped all over the place. It was an incredible ride . . . very personal. . . . There is the team aspect, of course. But yeah, it has been a personal journey. Very personal journey." Squyres felt overwhelming emotion when the rovers landed. "This is so good, I can't believe how good this feels. . . . Pancam is really on Mars after all these years. The whole damn thing is on Mars. I dissolve into tears."[17] He concludes his book on the MER missions with the comment: "I love *Spirit* and *Opportunity*."[18] He confesses the prejudice that one shouldn't say such things lightly about machines, but the emotion overwhelms him. These machines are "our surrogates, our robotic precursors to a world."[19]

The scientists involved in their MER missions identify with these rovers because like us, the rovers move, they sense, they scrape rock, move things around, take photographs, and send them home. For the MER scientists being a member of a mission team realizes a personal dream of being an explorer. Yet in the indirect world of telescience, these individuals who were trained in academic cultures that reward individual achievement must sacrifice a sense of personal agency and lose their individual voices. Projecting their identities into the robots helps to make this bearable. Everyone is equally anonymous, yet equally present in the robot, moving together, meters at a time across the Martian surface. For observers of the project, speaking of the rover as "discovering," is simple anthropomorphization. For the project scientists, it is a way to preserve identity.

Joined at the hip but advocates for their own particular disciplines, pursuing private scientific inquiries on the side, valued for their place on the team, but tolerating the submersion of identity because it is equally shared and because they are able to express themselves through the rover, this new technologically mediated science creates new scientists. Through the rover they forge a team that explores Mars as a group, coaxing their robots through treacherous Martian sand and steep rocky craters. Earlier technologies, some orbiting silently above a planet, some simply planted on a planet's surface, gave scientists stretches of time to consider and formulate hypotheses. MER with its two boots on the ground—moving along, shifting perspectives to confront new terrain—demands a different kind of thinking, a continually reoriented thinking-in-place. "Doing science" devolves into teleconferences, image manipulation, and computer analyses that provide their only contact with, yet further alienate the scientists from, the rocks and chemistry of Mars. Nonetheless, each person, in discovering a way of participating in relation to the robotic technology and its products, finds a unique place in a bigger picture.

MER importunes with its demands for high maintenance. It requires daily devotions of a particular sort. The scientists must attend personally day after day, year after year to the robot's next actions, and then attend again to the stream of data they have commanded. Such demanding devotion needs scientists with the imagination to organize a new "big picture" of discovery through the mediation of MER. The rover is the hero of this new narrative. In the epic of the rover the scientists write themselves back into the story of their own personal journeys, in what must be, for now, the remote exploration of our planetary system.

INTIMATE SENSING

Stefan Helmreich

We make much of the sea as an immersive medium. From Freud's "oceanic feeling"—a human nostalgia for a lost communion with a watery mother nature—to Jacques Cousteau's contemplative scuba diving and reverence for the underwater realm, European and American culture overflows with images of the sea as a zone in which boundaries between self and world dissolve.[1] And, indeed, some kind of full, unmediated presence in the deep sea has become a signal aspiration of oceanography. Marine scientists have moved from dredging the deep with buckets to mapping the seafloor with sonar to, today, employing remotely operated robots and human-occupied submersibles to deliver full-color portraits of the world below. These technologies do more than deliver information. As anthropologist Charles Goodwin has pointed out, marine scientists encounter the sea through thickets of technologies that come packaged with sets of social relations—relations often built right into the structure of research ships themselves.[2] At the center of my story here are relationships between marine scientists, ship's crew, and a deep-diving, remotely operated robot—known as an "ROV," for "remotely operated vehicle"—a device that offers tele-present optical access to the undersea world as well as the capacity to remotely move and manipulate objects in this realm. Although ROVs are not the sort of vehicles into which humans can physically crawl, many of the marine scientists who use them feel a direct body-to-body connection with these objects. These scientists see themselves as involved not so much in remote sensing as in *intimate sensing*.[3]

SEEING IN DEPTH

Over its century and a half of existence, deep-sea science has examined the ocean ever more closely, with the technical goal of *seeing into* the submarine realm. Oceanographers, as historian Sabine Höhler has suggested, have long deployed their tools with the ambition of approximating the visual, of creating images that evoke dimension and volume. "Depth," she writes, "became a matter of scientific definitions, of the right tools to see beyond the visible surface, of huge amounts of sounded data, and of their graphic representations. The opaque ocean was transformed into a scientifically sound oceanic *volume*."[4]

The ROV that is the focus of my narrative promises transparent visual access to the underwater zone by offering high resolution, full-color images of the seafloor and its life. It transmits images from the deep for real-time interaction with scientists and crew on ships. Unlike more everyday equipment oceanographers send down to get readings of temperature, salinity, and depth—equipment that delivers graphed readouts to ship computer screens—the ROV offers a more literal sensation of what Goodwin calls "seeing in depth."

Technologies of deep-sea telepresence alter a sense of self for both marine scientists and ship's crew. As visualization technologies "make possible new ways of seeing and doing," they also shape the identities—professional and personal—of those people who employ them.[5] Historian and ocean engineer David Mindell asks a question that sums up one concern of oceanographers using remote technologies: "Are you a real oceanographer if you don't descend to the seafloor? Are you a real explorer if you never set foot in a new world?"[6]

In what follows, I report on two oceanographic expeditions in California's Monterey Bay. Each day-long expedition was on the *Point Lobos,* a ship operated by the Monterey Bay Aquarium Research Institute (MBARI), the research sibling of the more famous Monterey Bay Aquarium, situated on California's central coast. My larger project examines the intersection of marine microbiology and genomics, in sites from Hawaii to Massachusetts. Here, I narrow my focus to what it is like to do science at sea in a computational culture, when research vessels are increasingly outfitted with robots, computers, and Internet connections. At least for the handful of marine biologists with whom I worked, ROVs are embraced as tools that make them feel as though they are "real explorers." ROVs provide a sense of close encounter but not necessarily that of a lone adventurer. Scientific experience of the deep sea via the ROV is mediated not only by its technological appendages—robotic arms and remote camera-eyes—but also by the skilled, collaborative labor of the crew who work on the ship and with the ROV more regularly than the scientists who cycle through.

For its part, the ROV exists within a constellation of controls and extensions—computer monitors and television screens, kilometer-long tethers into the sea—that can make it a very different beast to the range of actors involved in its deployment. Scientists and technicians at sea as well as computer programmers, video librarians, and Internet users on shore all develop distinct relationships with the ROV, relationships that constitute important aspects of their scientific and technical selves. The sensation of intimate sensing has a different feel for different players (and, indeed, may not adequately describe everyone's experience). The mediations are multiple and so are the selves.

THE SHIP AND THE ROBOT

I am groggy and it is early morning, 7 a.m. I am finding my feet on the drizzly deck of the *Point Lobos,* a small oceanographic vessel. Today I am joining the *Point Lobos* cruise as an anthropologist, doing a participant-observational study about marine biological fieldwork. The ship is gliding out of Moss Landing, a tiny town just north of Monterey.

This cruise will be a one-day expedition employing an ROV to examine deep-sea ecologies known as cold seeps, muddy seafloor environments that host methane-metabolizing microbes. These microbes keep toxic greenhouse gases out of our atmosphere. Seeps are ecosystems only recently discovered by marine science, thanks in large part to such technologies of access as research submersibles and robots. The chief scientist on board, marine biologist Rob Haldane, in his early thirties, briefed me a few days before our cruise.[7] We spoke over a precruise calamari lunch at Phil's Fish Market in Moss Landing. The task of the science party and crew, explained Haldane, will be to dredge up mud from methane-rich zones for further study; along the way, we will find tubeworms and clams that thrive in intimate symbioses with bacteria that live off of compounds poisonous to most creatures.

Every step of the way, our journey will be mediated—by water, camera, and computer. Some sensing equipment on the ship works without direct human guidance, but on our trip humans constitute a key part of the research ecosystem. There are practical reasons for this: guiding the robot demands human interpretations of what is important to investigate. There are also reasons to include humans that are linked to the culture of marine science: when scientists stay "in the loop," they preserve a sense that they are explorers. There are broader cultural reasons, too: because the deep is imagined as a mysterious place, a realm full of scientific surprise, scientists want to be there; indeed, without human apprehension, the sea would not manifest at all as a site of otherworldly difference. Oceanographic discovery is about *human* encounters with the sea. The cruise is not just about raw information; it is about *meaning*.

As we curve away from shore, the seven crewmembers of the *Point Lobos* hustle across well-worn paths, securing swinging doors and tying necessary knots down the length of our 110-foot ship. Today the *Point Lobos* carries a small party of researchers, including me. Haldane, who is coordinating the cruise, works as a postdoctoral fellow under MBARI microbial biologist Ralph Leiden, in his mid-forties, who has done pioneering work in environmental marine genomics, the sequencing of DNA from seawater. Haldane sets everyone at ease with his generous humor and quiet ability to cajole chaotic collections of ocean-going machinery into coordinated action. A graduate student from nearby University of California–Santa Cruz, Adam Trilling, in his late twenties, has signed on to participate in the sampling and will eventually be keen to figure out how to extract genetic material from the mud we collect. Nadine Rodin, also in her late twenties, is an assistant to an MBARI marine geologist. She has joined up to learn about the ship's global positioning navigation system.

I glance at a printout of our program for the day:

Cruise Plan, Friday, March 7, 2003
Ship: R/V Point Lobos
Chief Scientist: Rob Haldane

Subtract 7 hours during PDT to convert to local time, 8 hours during PST
Scheduled Start Time: Friday 7 March 2003 1500 UTC
Scheduled End Time: Saturday 8 March 2003 0030 UTC

Purpose: To collect seep sediments for dissolved gas analysis, nucleic acid extraction, and other analyses.

We are just getting out to sea when the drugs begin to take hold. Rodin tells me she has taken Meclizine. Trilling has downed six Dramamine and four Vivarin. Haldane says that he, too, is well prepared—though he jokes that he has steered clear of Scopolonia, which tends to make people hallucinate at sea, not ideal for doing careful marine science. I have taken nothing against seasickness. Having determined that everyone else is dosed up, I nervously consult my stomach.

We are summoned by Dave Wolin, a deckhand in his forties, who asks the four of us to sign waiver forms. He gives us a briefing on life jackets and lifeboats. Life, it would seem, is about floating. And floating at sea, Wolin tells us, is often about throwing up. He shows us the "place to yak," an area on the port side of the boat, a location, he kindly informs us, not monitored by the ship's cameras.

As Trilling and I sway about on deck, Rodin comes up from below and announces, "I just saw you on TV." *Point Lobos* is studded with cameras for two reasons. One is for safety, in case there is an accident. The other is to transmit images from the ship over to the Monterey Bay Aquarium, a major tourist destination. The gorgeous fish tanks at the aquarium invite patrons to "come closer and see." The ship's cameras allow patrons to come closer to us. What we see on the ship will become a distributed experience.

The most important cameras on the *Point Lobos* are not attached directly to the ship itself. They are built into the massive, two-and-a-half-ton robot sitting on deck. This is *Ventana,* the ship's ROV, which can be dispatched into deep water off the ship's port side. The half-a-million-dollar *Ventana* uses space-age technology to withstand extreme atmospheric pressure and temperature.

Ventana is Spanish for "window," and, true to its name, it offers a framed glimpse into the ocean. It receives commands through a stream of fiber-optic cables running from the ship—what researchers call an "umbilical cord." This tether allows the robot to travel down to 1,500 meters, a zone characterized by crushing pressures and permanent darkness. This the *Ventana* can illuminate with full-spectrum and incandescent lamps. The images *Ventana* captures from the depths are transmitted up to the ship's control room. There, they can be monitored in real time and, if desired, sent via microwave to shore and uploaded to the Internet. On the Internet a curious public can leap into the virtual deep.[8] The robot is outfitted, too, with manipulator arms and a suction sampler to collect tangible items like starfish and clams.

The *Ventana* is itself a condensed history of deep-sea exploration. During the earliest oceanographic voyages, in the late nineteenth century, ships like Britain's HMS *Challenger* drew their knowledge of the sea from dredging—bringing up objects from the sea bottom using bags or buckets attached to piano wire. *Ventana*'s grasping metal manipulators will later today deliver coarse, cold mud up from the seafloor, sediment not very different from that caressed by Victorian naturalists. Sonar, or *sound navigation ranging*, an invention of the early twentieth century created to detect submarines, is used to help steer the robot. *Ventana*'s prominent zoom lens, attached to a Sony HDC-750A high definition camera, plugs us into the history of innovation in underwater photography, which has now definitively entered the digital age.

The use of computer technology to sound the sea is a central goal of MBARI, founded by computer entrepreneur David Packard, a cofounder of Hewlett-Packard. MBARI has become a key player in the development of deep-sea robotics as well as the development of telepresence techniques for ocean ecosystems research. In recent years, MBARI researchers have also been drawn to genomics, sequencing DNA from sea creatures such as those we hope to gather today.

Ventana has been outfitted by instrumentation technician R. L. Hopper with a bank of plastic cylinders that will be pushed into the seafloor to collect cross-sections of methane-infused ooze. Looking for signs of life at the bottom of the ocean is a relatively recent possibility. In the early nineteenth century, naturalists thought the deep to be devoid of life, in part because of a prevailing belief that seawater was *compressible,* that

seawater grew more and more solid until a point was reached beyond which a sinking object would sink no farther. Thus, somewhere in the middle regions of the great abyss, there existed "floors" on which objects gathered according to their weight. Cannon, anchor, and barrels of nails would sink lower than wooden ships, which in turn would lie beneath drowned sailors.[9]

The ocean into which *Ventana* descends is much transformed from the ocean of the Victorians. We now know it to be a medium with living things all the way down.

As I look at *Ventana's* plastic cylinders, called *push cores,* Trilling talks to me about marine mud. He points out that methane oxidation by methane metabolizing microbes is environmentally important because almost 75 percent of methane emanating from seeps and other methane systems may be consumed by these bugs. If it were not, he laughs, we humans would be drowning in smelly methane. We head inside to the mess, where we study a wall map of the Monterey Canyon, on which serpentine lines representing the underwater canyon's branches snake out from a shore-based origination point at Moss Landing. This sonar atlas delivers a from-the-sky blueprint of a topography invisible to the eye. It is a perfect symbol of the transparency promised by *Ventana,* a digitally fashioned dreamscape in which imaging media erase their work of mediation.

We arrive at our dive site, twelve nautical miles out of Moss Landing, 36.78°N, 122.08°W. We are not that far from land. The Monterey Canyon, deeper than the Grand Canyon, escorts deep water close to shore. We settle into a one-hour wait until the robot reaches the methane-saturated ecologies that interest Haldane. The color of the water on the ship's TV feed from *Ventana* begins to change, from a light blue to a thick azure, and finally to black.

ENCOUNTERING AN EXTREME MARINE WORLD

Ventana has arrived at the seafloor and now floats just off the bottom, one thousand meters below the ship. Scientists on board congregate in the ROV control room, a snug, wedge-shaped chamber squeezed into the front of the ship, on the lowest deck. In the dim light, we can almost discern the inside-out outline of the bow. The *Point Lobos* used to be an oil-field supply vessel named the *Lolita Chouest*, and this room housed sailors' sleeping quarters. Where once people sunk into a seesaw sleep, scientists now remotely ride the *Ventana*.

Ventana's cameras point in many directions. We are most interested in the forward view, presented on several screens, at many sizes in the ROV control room. The most prominent video screen is adjacent to a VCR, just next to a monitor hosting an annotation system called VICKI, for "Video Information Capture with Knowledge Inferencing." Next to the VICKI is a monitor displaying "frame-grabs" from ongoing ROV video feed; frames are captured when a researcher clicks the appropriate icon in the annotation system. On this trip, at least at the beginning, that researcher will be me.

Haldane has stationed me in a chair in front of VICKI, where I will change videotapes every hour and click pictures as instructed. He hands me a timer. As I look around the room, each screen sways to its own rhythm. Haldane seats himself to my left, in front of a keyboard from which he can control *Ventana's* cameras and lights.

The ROV pilots, Jerry Malmet and R. M. Engel, sit in the leftmost two chairs, which are outfitted with joysticks for steering, or, as they prefer to say, *flying* the ROV. The pilots also control two robot arms, with which *Ventana* can be made to nudge things, grasp instruments, and pick up stuff. Rodin is sitting in back, studying the navigation console. We are all wearing headsets, allowing us to speak with one another over the dull, slowly modulating hum of the boat. Haldane addresses me through my headset: "Stefan, go ahead and snap as many frame grabs as you want of the push cores going in, coming out, whatever looks good." A CD player has been left on and quietly percolates a gentle reggae. The room is a multimedia event and a sensory scramble, a layering of ocular, auditory, and corporeal disorientations. We project our presences into *Ventana,* whose cameras are now to be our steady eyes.

Haldane instructs me to take frame grabs of tubeworms that we have found near a methane seep. Living inside these tubeworms are microbes that metabolize sulfide. We are far from the sunny, salutary ocean. One biologist at MBARI suggested to me that rather than demonstrating that the ocean is a zone of ecological harmony, sites like seeps should lead us to see the ocean as a giant refuse heap; life exists here not because the medium is so friendly to life, but because life is so adaptable, and can make its way in the most noxious of environments. I am not sure if this is fair to tubeworms, though it is true that they have been persistently associated with the creepy, unearthly, even extraterrestrial—a chain of associations more emotional than logical. But the alien connection is a prevalent one, even for some scientists. Haldane tells me that enthusiastic artists have plopped tubeworms into fancifully painted seascapes of alien planets, a graft he says makes no sense. If metazoans evolved on other worlds it is unlikely they would so closely resemble earth creatures.

Science fiction turns out to be one of Haldane's inspirations for becoming a deep-sea biologist. Over our lunch in Moss Landing, he told me that as a child he was riveted by *Star Trek,* by the notion of traveling and doing science in three dimensions (Haldane's mother, not insignificantly, was a ground engineer for some of the *Apollo* missions to the moon). As an adolescent, he studied the migration of large animals and particularly liked the movie *Star Trek IV,* in which the crew of the *Enterprise* returns to twentieth-century Earth to save the whales (specimens of which they find at a fictional "Cetacean Institute" portrayed in the film by the Monterey Bay Aquarium!). The similarity between the view-screen on the starship *Enterprise* and the screens through which we look at the video feed from *Ventana* makes the comparison with *Star Trek* seem natural. Some of Haldane's older colleagues grumble that MBARI has not invested in human-occupied submersibles, but Haldane so identifies with the ROV—it feels to him an extension of self—that he disagrees. For Haldane and for his counterpart on the next cruise I would join, ROVs provide a satisfying sense of being there, of immersion. Goodwin observes of abstract, computerized graphs of deep-sea data fed to ships something that might actually apply more fittingly to images on ROV video screens: "Like the screen in a cinema, these inscriptions are the focus of intense, engrossing scrutiny. Indeed, they are the place in this lab where phenomena in the world the scientists are trying to study, the sea under their ship, are made visible."[10]

The scientists I study *want* to be immersed in the sea. In some sense, they want to lose themselves in the place they study. Many grew up on the beach and know how to scuba dive. They love the ocean. The story of MBARI science and ROV *Ventana* would be very different if the scientists working with these tools found the ocean to be terrifying and monstrous, a view that predominated among most people before the rise of nineteenth-century Romantic visions of the oceanic sublime. In the twentieth century, this sublime fused with ecological imaginaries of the beautiful, fragile sea. People came to see the union of self and sea as one of the most privileged ways to appreciate nature.[11] The techno-immersion offered by the ROV builds on this sensibility.

In the instance of oceanographic research Goodwin observed, scientists and crew occupied different locations on the ship, controlling different aspects of the same device from separate indoor and outdoor locations. They existed as if in different worlds. Goodwin suggested that that research ship embodied "a historically constituted *architecture for perception*." Oceangoing scientists and crew enacted "not simply a division of labor, but a division of perception."[12] This is true with ROV operation, too, though to a less extreme extent.[13] Although there are mechanical aspects of managing the *Ventana* that scientists would be ill prepared to undertake, operations in the ROV control room put the scientists and crew in close relations of interdependence.[14] It is understood that scientists direct the action, but the boundaries are often blurred. An exchange between Haldane and ROV pilot Malmet illustrates. Note the shifting use of pronouns:

Haldane: Tubeworms! We landed right on them. Can I zoom in a little bit? Look at all of those! Let's do a flyby. Is it easy to get some of those tubeworms?
Malmet: We can't guarantee that we'll be back to the tubeworms, so let's do it now.

Haldane: Stop here. Come wide. Zoom in. There's one right next to your wrist there.

Both ROV pilot and scientist use "we" to refer to the joint enterprise of maneuvering the robot. Haldane slides into "I" to flag his particular concern and "you" to direct the ROV pilot's embodied expertise. When he says "your wrist" however, he is referring not to the pilot's fleshy arm but rather to a joint near *Ventana*'s claw. The substitution is telling. I am told that the ROV control system makes use of software similar to that of the Sony PlayStation. The ROV uses joystick controls familiar to those who have played video games on this platform. Video games are opaque, the innards of their programs hidden from users—a fact that, paradoxically, people experience as "transparency."[15] A similar dynamic is at work with the ROV. Its opacity/transparency makes it possible to identify one's hand with the claw, as in a "first-person" video game. Like any prosthetic technology, the system affords scientists with "perceptual access to the world they are sampling, while simultaneously shaping what they are able to see there."[16]

In her analysis of jellyfish tanks at the Monterey Bay Aquarium—exhibits crafted to transfigure viewers by bringing them into intimate relation with seductively displayed cnidarians—historian Eva Shawn Hayward says that, "Water is the proposed factor in the refiguring, but the viewer is endlessly confronted by the various instruments that produce the experience. Immersion, then, here, conveys the experience of being totally inside the technologies and ecologies of this MarineWorld."[17] Similarly on the *Point Lobos*, the sensation is not of detachment from nature, but of a pleasurable, technological immersion *in* it—an experience felt as at once immediate and hypermediated (that is to say, mediated in lots of ways, all at once).

By now, a whole bush of tubeworms has been pulled free of its moorings—a process I have been documenting with frame grabs. (Later, I will visit the Institute's video lab, where such images are annotated. Since 1988, MBARI has made some 12,000 tapes of dives, in assorted formats. At the institute onshore, three women work in the video lab, a windowless, temperature-controlled room. They review footage and mark when and where creatures show up on the tapes. One of the video librarians calls bemused attention to the gendered division of land-and-sea labor that has women like her running between kids and work, while an all-male crew of ROV pilots stays out all day operating heavy machinery.)

We deploy another push core. It presses into the mud and captures a clam. "This guy comes back with us!" says Haldane. We spy sessile relatives of jellyfish. I glance at my monitor and see that it features a digital count of Greenwich Mean Time, tweaked to display melting numerals. The numerals morph, Dalíesque, into their successors. Surrealist invention is allowed to flower in the corners of our computer screens, but realism is sternly enforced for the screens delivering images from *Ventana*. Haldane continues in movie-director mode, panning and tilting the ROV amidst the ship's pitch, rock, and roll.

I realize that my stomach, too, is rolling. I get up to find some air. I race to the yak zone. Other people in the science party have gotten sick, too, in spite of the drugs. Trilling remarks that it is hard to do science under these conditions; you're either sleepy or you're sick. And you do not want to throw up all over the control room.

I remark that this is not the meditative sea so celebrated in poetry, but more like the perfect storm in your stomach.

When I reenter the control room, Trilling, without comment, has done my job of changing the videotapes and taken my place at the screen. I meditate on the floor at the back of the room. Intimate sensing turns out to depend not just on the opacity/transparency of technology, but also on seasickness pills! After a while, operations are finished and the ROV begins its slow ascent to the surface. People leave the room to get fresh air.

DIVERGENT PERSPECTIVES ON THE INNER HISTORY OF THE ROV

On a second *Point Lobos* cruise, on which we focus on the study of symbiotic bacteria that reside inside marine invertebrates, the social dynamics in the control room are much the same as they were on the cruise with Haldane. On this cruise, the director is Tamara Robena, in her mid-thirties, a MBARI specialist in the clams and tubeworms that live at methane seeps.[18] At ease and experienced, Robena gazes into the cathode ray tube transmitting the ROV's point of view. She asks ROV pilot Thad Cormant, mid-forties, to turn this way and that, to poke at the mud. Clam beds come into view. Robena tells the pilots where to place ROV's claws and how to position its push cores. A push core is drawn from the ROV's quiver of twelve and the question arises of whether more than one sample should be packed into the same tube:

Cormant: How about multiple pushes with the same core? Probably fine if you don't want context. We're not with geologists today! Hmm. This push isn't working.

Robena [in pirate voice]:   That thar be rock. Hey! That looks like a bacterial mat!

Cormant pulls back. Robena moves the camera using a mouse pad. We find more clams.

Robena: There's a clamshell already. Clam-o-rama!
Cormant: Scoop them up? I'll just drop them in the drawer. [*Ventana* has a bin for storing things picked up on a dive.]
Robena: Slurp those guys!

The ROV pilot projects himself into the body of *Ventana,* transposing his movements into the robot below. He is gingerly trying to keep the robot from making too much commotion but is not being entirely successful.

Cormant: Might be some collateral damage, but it's okay. Hey, what's that?
Robena: Well, I have no idea. That could be a sponge, but maybe bacteria. Can we get it? [He imitates a radio transmission from a bad science fiction film.] We think it's a blob. Nope, it's sliding out. No blob for us.

Robena and the ROV pilots speak as though they are connected—almost as if they share a body, with Robena the eyes and the pilots, in the instance here, acting as the hands. As if to reinforce this overlapping division of perception and labor, Robena refers to herself as "I" when she is talking about ideas and "we" when she and the pilots relate through the ROV.

I ask Cormant how he maneuvers *Ventana.* He demonstrates its robot arms. The left one has several joints, named with words suggesting an amalgam of the mechanical (swing), the bodily (shoulder, elbow, wrist, grip), and the nautical (pitch and yaw). The right arm, which is newer, boasts hydraulics capable of lifting 500 pounds. The user can freeze the arm's position in computer memory and then, after moving it physically, reactivate the stored position, which now no longer corresponds to where the arm is in real space. This is useful if you get your own fleshy arm into a contorted position. Cormant lets me play with *Ventana*'s arm as the robot rises to the surface. It is an experience of disconnection, if not disembodiment; there is no force feedback (he tells me) from the joystick. This reinforces the sense of weightless, outer-space-like travel that inspires comparisons between the ROV and space ships.[19] This is telecommunication without teletacticality, intimacy without immediacy, a gap that makes explicit the work required to realize *telepresence*, presence from afar.

The seemingly seamless experience of intimate sensing—individual and communal—is not only enabled by the different yet interdependent perspectives of scientists and crew working together. It must be actively maintained, supported. Simply put, there is a fair amount of work necessary to keep the ROV running. Cormant tells me that sometimes he is up until midnight fixing things before a dive day. He often serves as a mechanic to the ROV; it is like a car that he has to look after and fix. He sees into it in a way that the scientists do not, and it is important for him to have a sense of mastery over the device. He gives me an example of a modification that the ROV pilots made to *Ventana*. Normally, the *Lobos* finds out where *Ventana* is through "pinging" a sonar signal off of it. The information is later fed to shore, where computer programmers receiving data points use a mathematical formula to come up with a best-fit line for the path of the ROV for a particular dive. The pilots, however, have not found the sonar as fine-tuned as they would like. So, they have outfitted the ROV with a Doppler Velocity Log, measuring movement relative to seafloor (rather than in "absolute" space). This has been accurate about how *fast* the ROV is going, but not always about *where* it is. When the data from this modified ROV system has been fed to shore, software engineers have found themselves confused about the absolute coordinates of the robot, since they were not initially privy to the modifications—done in the name of shoring up ROV pilots' experience of control. But Cormant mentions that sometimes the software people confuse the people at sea, too, for example when they upgrade the ship's software without notice. Sometimes the onshore technicians even make modifications in programs during a sail: cursors move around on ship screens as though the boat is haunted. Between those at sea and those on land there is a difference in perspective and of who is authorized to exercise power where and when. The disjunctures that result can unravel people's sensations of intimate sensing.

Shoreside software engineer Jay Bluestone, in his mid-forties, is one of the people who had to grapple with the mysterious Doppler modifications. His job has been to put together a relational database from the many kinds of data collected by *Ventana*. My interview with him centered on how he thought about the sea. For Bluestone the ocean on the screen was conceptual, not immediate. Dealing with traces of the ROV's path, and with data it delivered, he was not able to project his consciousness into the robot. When he considered the ocean through the medium of his database, he said, "I don't think of it as a wet thing. It's a construct that places constraints on what we do. It's not the same ocean that we go to when we step outside and go to the beach." For him the ocean is alien: "*The ocean is not of us.*"

As Cormant and I watch the screen in the *Point Lobos* control room, a docent at the Aquarium interrupts our conversation. She appears as a tiny image on one of the video screens. She says in a tinny TV voice:

Nothing is larger or more important than the ocean. We get food, minerals, and pharmaceuticals from it. And with the ROVs, we can enter this alien world. None of this would be possible without MBARI's research vessels. The scientists are on the boat right now! They can beam images to us live. If I go live to the boat, we see . . . not much.

She speaks to us for the benefit of Museum visitors: "*Point Lobos,* are you there?" she asks. Cormant says yes. "What did you do today?" Cormant explains about the clams and the tubeworms and the cold seeps. The docent then proceeds, on the fly, with the help of a vast video menu, to explain to her audience what all these are. "Over and out," she says.

We all end up on the bridge of the *Lobos.* There is talk of decommissioning the ship, transforming it into a craft that can maintain elements of MBARI's next technological undertaking: a distributed ocean observatory, a network of remote sensing buoys that can provide continual Web access to data from the sea. I later hear a talk at MBARI about how such networks would allow scientists to sit in their living rooms gathering oceanographic data. No need for seasickness pills or yak sites. The *Point Lobos* might do maintenance in such a system (and *Ventana* might shift from a glamorous tool for exploration to an everyday repair device, rearranging technologies and selves yet again). A crewmember jokes that without scientists onboard, they could do some proper salmon fishing. Extending associations between the oceans and outer space, projected marine observatories have acronyms like MARS (Monterey Accelerated Research System), NEPTUNE (NorthEast Pacific Time-series Undersea Networked Experiments), and VENUS (Victoria

Experimental Network Under the Sea). In the future, scientists might explore the ocean by surfing the Web. One scientist who looked forward to marine observatories told me that when they came online, she would be able "to bring the ocean into [her] living room." This will be a new order of intimate sensing.

THE MULTIMEDIA OCEAN

Marine biologists at MBARI encounter the sea as a media experience, one in which they seek to be experientially and technologically immersed. The ocean encountered by researchers using *Ventana* is the object of a rapt optical attention aimed at comprehension through vision. Given the mission of MBARI, it is not surprising that the deep appears as a place to be technologically explored, a darkness to be illuminated. This underwater domain oscillates between the unfamiliar and the ready-to-be-apprehended, if not domesticated. We might find, as the MBARI software designer told me, that, "the ocean is not of us," that the ocean represents, as the Aquarium docent put it, an "alien world." Or, we might hear—as I did when I was certified as a scuba diver just south of Moss Landing—that, "The ocean really is us. We are its eyes." My dive instructor meant to enlist his listeners into an ethos of ecological monitoring, into an intimate enfolding in which we could alternately appreciate oceanic difference and commune or identify with it.

Much of such rhetorical oscillation, such fashionings of intimate sensing as at once elusive and achievable, has to do with apprehending the ocean as a medium characterized by its shifting transparency and opacity. Wavelengths of light taper off into blackness the deeper one travels. Descriptions of the deep as dark and *therefore* mysterious, full of secrets, unknown, draw on a reservoir of association linking sight and light with knowledge; indeed, the word *theory* derives from the ancient Greek *theorein,* which means both "to see" and "to know." Visualizing the ocean has become the governing goal of oceanography, the grail of techniques of remote sensing.[20] With ROVs, such visions are imagined as at once immediate and objective.

In recent years, the objectivity associated with vision has hybridized with computer imaging and the oceans are now coming to be viewed through online interfaces, or through the robot cameras of entities like *Ventana.* Oceanographic knowledge is nowadays uploaded into computer images and text. Understanding today's scientific sea means engaging with this new media ecosystem. Marshall McLuhan once suggested that "the medium is the message"—that is, that media extend and modulate our sensorium, that "the effects of technology do not occur at the level of opinions or concepts, but alter sense ratios or patterns of perception."[21] With this lesson in mind, we can say that the medium of water, now fused with technological media, offers a new kind of immersion, an intimate sensing of a multimedia ocean.

In *The Second Self,* Sherry Turkle argued that in the early days of the personal computer, people came to see the machine as a kind of mirror for their own minds—a second self. In these days of distributed, polyglot computing and visualization of the sort crystallized by *Ventana,* people see constellations of computers not so much as second selves but as an array of selves that move along a number line from the zero point of self-identification to the multiple identities of distributed, prosthetic subjectivity. Not just life on the screen (to borrow the title of another Turkle book), then, but lives on, through, with, and between the screens.[22]

Buildings and Biology

KEEPERS OF THE GEOMETRY

Yanni A. Loukissas

"Why do we have to change? We've been building buildings for years without CATIA?"[1] Roger Norfleet, a practicing architect in his thirties poses this question to Tim Quix, a generation older and an expert in CATIA, a computer-aided design tool developed by Dassault Systemes in the early 1980s for use by aerospace engineers.[2] It is 2005 and CATIA has just come into use at Paul Morris Associates, the thirty-person architecture firm in the Southwest where Norfleet works; he is struggling with what it will mean for him, for his firm, for his profession. Computer-aided design is about creativity, but also about jurisdiction, about who controls the design process.

Architectural theorist Dana Cuff writes that each generation of architects is educated to understand what constitutes a creative act and who in the system of their profession is empowered to put themselves in a creative role.[3] Creativity is socially constructed and Norfleet is coming of age as an architect in a time of technological but also social transition. He must come to terms with the increasingly complex computer-aided design tools that have changed both creativity and the rules by which it can operate.

Traditionally, architecture has been defined by its practitioners and patrons in relation to three sets of standards: technical, economic, and aesthetic. Buildings must be sound, practical, and beautiful. Vitruvius, author of one of the earliest known architectural treatises, *De Architectura,* expressed these qualities in Latin as *firmitas, utilitas,* and *venustas.* Modern architects have maintained distance from technical and economic activities in order to privilege their aesthetic role, what the sociologist Magali Sarfatti Larson has called the "aesthetics of construction."[4] However, with new technologies of simulation, embodied by programs such as CATIA, things are changing; new forms of technical expertise are becoming central to the architect's professional identity.

In today's practices, architects use computer-aided design software to produce 3D geometric models. Sometimes they use off-the-shelf commercial software like CATIA, sometimes they customize this software through plug-ins and macros, sometimes they work with software that they have themselves programmed. And yet, conforming to Larson's ideas that they claim the higher ground by identifying with art and not with science, contemporary architects do not often use the term "simulation." Rather, they have held onto traditional terms such as "modeling" to describe the buzz of new activity with digital technology. But whether or not they use the term, simulation is creating new architectural identities and transforming relationships among a range of design collaborators: masters and apprentices, students and teachers, technical experts and virtuoso programmers. These days, constructing an identity as an architect requires that one define oneself in relation to simulation.[5] Here, case studies, primarily from two architectural firms, illustrate the transformation of traditional relationships—in particular that of master and apprentice—and the emergence of new roles, including a new professional identity, "keeper of the geometry," defined by a fusion of person and machine.[6]

PAUL MORRIS ASSOCIATES

Little more than a year ago, Paul Morris hired Quix to teach architects how CATIA might be used to rigorously model their designs in three dimensions. Although Quix was never trained in architecture, he has been working in the field for almost two decades. Quix once taught and sold CATIA to its intended users in the aerospace industry, working primarily as an employee of IBM, the parent company of Dassault Systèmes. Now in middle age, Quix is using his expertise in CATIA to create a role for himself in architecture, a field he flirted with briefly as a young man, but turned away from to pursue his fascination with computer-aided design. Quix is teaching CATIA at Paul Morris Associates and also working to apply the software to the firm's current projects. At Paul Morris, there is active resistance to his presence. Control over CATIA translates into a great deal of control indeed. Before designs are built, they only exist as representations; whoever models a project produces and controls the current reality of the design.

At Paul Morris Associates, resistance to CATIA is shaped by many considerations. Rikle Shales, an architect in her mid-thirties, resists putting designs into CATIA because she says that once they are thus represented, people tend to see the designs as frozen, as a done deal. She points to the number of design changes that take place in an ongoing project and argues for keeping designs out of CATIA until late in the process. She says, "Quix has a habit of thinking that if we just put things into CATIA, they will be done and coordinated. It's pointless to model the current design in CATIA when it is almost certain to change." Her position is well reasoned, but she and the architects around her know that it has a political dimension. Keeping designs off CATIA lets her stay in greater control of the design process. Once within CATIA, one might say that all designs belong to Quix because he is the one best positioned to manipulate them.

Quix and his architect opponents define themselves in relation to CATIA and one another. From Quix's perspective, architecture needs CATIA, which can bring a new rigor to building. He reasons that architects have difficulty learning CATIA because it requires a "level of rigor that architects are not used to." His colleagues' resistance to CATIA reinforces his sense of being different from those around him; he is the one who is a rigorous engineer.

Indeed, Quix describes his life at Paul Morris Associates in terms of "three difficult phases," each a stage that includes some resistance to CATIA. He calls the first phase "the brick wall." In this phase, he says, architects complain that they are too busy to learn. They say that the program is foreign; it feels like somebody else's approach to architecture. Quix refers to this first phase as "a sickness called NIH (Not Invented Here)." In the brick wall phase, Quix is isolated.

A second phase begins when Paul Morris, the founding principal of the firm, actually orders a team of architects to learn CATIA, with Quix as their tutor. Quix teaches the practitioners one by one.

In this second phase, tutelage with resistance, CATIA is viewed very differently by the practitioners and their teacher. Quix sees CATIA as a new way of looking at the world, a systematic and three-dimensional way of approaching design. Most centrally, it is a new way of being an architect. Quix feels he is bringing more than a new piece of software to Paul Morris Associates. He is bringing its architects a new epistemology and a new identity.

For the designers who are his students, CATIA is simply a technique, one more skill set to apply to their practical problems. The conflict between Quix and the architects of Paul Morris Associates calls to mind Sherry Turkle's description of two ways in which technology can be "transparent."[7] A "modernist" transparency enables a user to gain access to the inside workings of a system. It evokes the aesthetic of early relationships with cars in which one could "open the hood and see inside." This is the kind of relationship to CATIA that Quix desires. Turkle contrasts this with the "Macintosh meaning" of the word transparency. This is a transparency that reverses the traditional definition. It says that something is transparent not if you know how to make it work but if you can use it without knowing how it works. It is the transparency of the user who navigates the surface of the system, but does not have access to underlying mechanisms. Its aesthetics are postmodern. This is the understanding of technology that interests many of Quix's students.

Robert Laird, an architect in his early thirties, is someone Quix considers something of a protégé. Laird has mastered working on the "surface" of CATIA. Laird has been using 3D modeling programs since his years as an undergraduate majoring in architecture, AutoCAD, 3D Studio MAX, Rhinoceros, Form-Z, and has complaints about all of these platforms. For example, he says,

"Rhino is like 3D for dummies." "Form-Z produces gaps or leaks." Laird sees CATIA as the system that finally gets things "right." "It does everything you need." Laird puts confidence in CATIA without a deep technical understanding of how it works. "CATIA gives me the feeling that I have control when I use it. . . . Other systems will only get to a certain point before crashing, not CATIA." Laird claims that he was hired at Paul Morris Associates because of his skill with computer-aided design. Often, he works exclusively on the computer or coordinates the computer-aided design work done by others, a role he and his peers call being the "keeper of the geometry."

Laird's professional identity is constructed around his relationship with computer-aided design. Although Laird has become Quix's best student with CATIA, he continues to see the software in his own way. Quix teaches CATIA as a means of establishing technical control over designs. However, in Laird's hands, CATIA "feels artistic." For Laird, CATIA is a medium "for the manipulation of light and shadow."

A third phase of Quix's teaching is comprised of implementation with continuing resistance. In phase three, designers at Paul Morris use CATIA (under Quix's tutelage) for the design of a large project, a major public building. By this stage, the designers feel that using CATIA makes them part of the evolving direction of the firm. Yet, even at this point, resistance to CATIA continues, although it takes new forms. For example, despite his affection for CATIA, Laird has reservations about its practicality. He complains that it is not architects but consultants and contractors who are its main beneficiaries: "It is focused on making someone *else's* job easier." Laird is concerned that at Paul Morris Associates, people spend too much time modeling buildings on computers. "We are dumping hours in modeling." Laird's position is echoed by a senior designer who says that computer-aided design increases architects' tendency to "fetishize drawing" and create too many details too soon in the design process. Even with CATIA's use mandated by the head of the firm, Quix is still on the defensive.

Indeed, within the firm, Rikle Shales has carved out her identity by avoiding CATIA altogether. In architecture school, Shales was something of a computer guru. She was the "go-to person" for other students. But in Paul Morris's office, Shales keeps her knowledge of modeling to herself. When she works on the computer, she sticks to 2D programs; mostly, she communicates her ideas through sketches. Morris specifically asked her to learn CATIA, but Shales never got around to it, something she explains as due to the weight of other responsibilities. Shales voices some regret about not learning the program. She says, "I like to feel that I am keeping up with everything," but in fact, Shales's authority over the public building project that "lives" on CATIA has been due, at least in part, to the fact that she has not learned the system. The time she has saved has freed her up to do project administration. Beyond this, not knowing CATIA has confirmed her in a nontechnical identity; she is seen as a people person and design person. Others describe her as the "glue" that holds projects together, organizes tasks within the firm, and coordinates communication with outside consultants.

While Quix and Laird forged their roles within the firm by identifying themselves with technology, Shales made her place by keeping it at arm's length. Paul Morris, the head of the firm, has done neither. He does not represent himself as a technical expert but has allowed the technology to be an active player in his relationships with colleagues, most notably his relationship with a younger generation.

When the first American architectural practices emerged in the nineteenth century, the master architect was an active participant in every aspect of his office's work.[8] Theorists Donald Schön and Dana Cuff write about the strong master-apprentice relationship that developed in this context. Each argues, in their own way, that design is often best understood through a relationship between individuals, the master and apprentice. Dana Cuff writes that this is the "principal social relation" in architecture.[9] Schön describes traditional design education through the intimate involvement of a master architect, Quist, in all the particulars of Petra, his student's, work. Schön illustrates how closely they work together, with Quist's hands on Petra's drawings:

Quist places a sheet of tracing paper over Petra's sketches and begins to draw over her drawing. As he draws he talks. He says, for example, "the kindergarten might go here . . . then you might carry the gallery level through—and look down into here . . ."[10]

At Paul Morris Associates, computer-aided design software is reconfiguring the master-apprentice relationship. Paul Morris is a prolific practitioner and retired academic in his late sixties who gained early acclaim for his striking modernist buildings. He belongs to a generation of which architectural theorist Reyner Banham writes, "Being unable to think without drawing had become the true mark of one fully socialized into the profession of architecture."[11] Morris still relies on a pencil to develop new ideas. However, within his office, computer modeling is replacing many of the traditional tasks of the pencil, such as drafting and rendering. And although Morris rarely touches the machine, it has transformed his personal practice as well. He often works on the concept phase of designs with an apprentice, a younger employee who can help him model a range of design ideas.

Drew Thorndike is one of the apprentices who work with Morris to prepare sketches for design competitions. Thorndike has worked with computer-aided design since he was in college and continues to spend time learning new software. When asked about how design software figures in his long-term goals, Thorndike replies eagerly that they play "a big part. . . . I want to stay ahead of the game. . . . Computers equal speed and if I don't learn the new software, I will be out of the market. I am convinced of that."

From their first interactions on a project, a pattern emerges: Morris makes a sketch and Thorndike translates the sketch into a geometric model on the computer. Morris always asks Thorndike to print out the computer images he has generated. Then, Morris and Thorndike sit together with a roll of tracing paper and, in a close variant to the practice of Quist and Petra that Schön writes about, Morris draws revisions over Thorndike's prints.

Over time, Morris stops requesting prints and begins to look directly at the images on the computer. Now it is not unusual to hear Morris say to Thorndike, "Let's go look at your computer." The concept sketches are still drawn by Morris, but as projects progress, he and Thorndike work together at the computer; Thorndike operates the machine while Morris gives him guidance. From Thorndike's point of view, his boss has "learned to accept the technology on his own terms." From the point of view of other employees, Morris has conceded an important role, or as one puts it, "He is no longer the one 'sculpting space.'" But Morris has maintained what was most important to him. He wants to stay in touch with the evolving model no matter where it is, even if he has to sacrifice some of the physical intimacy he has with pencil drawings.

As they work, Morris and Thorndike are looking at the same screen, but they are not seeing the same thing. For example, when Morris makes a sketch for a new public library, Thorndike's job at the computer is to translate it into a modeling program called Rhinoceros, but which he affectionately refers to as "Rhino." Thorndike usually sits at the computer while Morris stands behind him. Rhino presents them with the virtual equivalent of a blank sheet of paper, a faint Cartesian grid floating in boundless space. With a click of the mouse, Thorndike chooses a starting point for an arc defined by three points on its circumference. Next, he chooses a section of an ellipse. With these two design elements, Thorndike generates a curved surface for the library's façade. Thorndike is choosing from a predefined set of shapes that the computer knows. These are his "primitives." Morris is watching his design emerge; he does not see into Rhino's inner world.

Rhino produces each of its images by simulating thousands of rays of light that reflect off primitives in the model and penetrate a virtual "camera," a virtual eye. This technique, known as "ray tracing," is based on an early Renaissance mechanism for transferring 3D objects onto a flat surface. So, for example, when Thorndike rotates the screen model, Rhino actually computes a new camera position for the ray tracing algorithm. Thorndike understands how the image is being produced. Morris sees only the rotation and the final image.

Morris is aware of the principles that drive Rhino's renderings. But when they talk about what is on the screen, Thorndike can make reference to details of Rhino's primitives and the ray tracing algorithm. Morris can talk only about lines, surfaces, and colors—at best a snapshot of the design. Before the technology intervened, Morris worked with images on paper, which he could physically manipulate. Now, his apprentice has a more direct experience of the developing design. Morris's experience is mediated, dependent.

Maurice Merleau-Ponty suggests that people incorporate instruments into their physical sensibilities through the experience of manipulating them.[12] From this point of view, Morris and Thorndike have a different experience and therefore a different understanding of the model. Morris's is more distant, Thorndike's more embodied and internal.

Thorndike is proud of his expertise with Rhino and other modeling software. He says, "I've taken AutoCAD to the limit. People in the office are often surprised by what I can do with the program." Yet he, like his mentor, still thinks with a pencil. Thorndike describes the evolving design for the library that he is modeling under the guidance of Morris: "The room is an elongated almond. . . . The entrance is tight and tall and the center of the space is wide but the same height." As Thorndike speaks, he takes a thick dark pencil out of his pocket and draws the profile of the library onto the bare surface of one of the drafting tables. His identity as an architect, like that of his master, lies between technological worlds.

When Thorndike becomes frustrated with Rhinoceros, when he reaches a limit, he turns to other modeling solutions. He is currently producing a new model for the library with AutoCAD. The AutoCAD model has qualities like repetition and flat surfaces that make it easier to build, both virtually and physically. Thorndike explains that the warped surfaces used in Rhino would be very expensive to build, especially in glass. In order to model the library in AutoCAD, the design had to be simplified, recast in a "platonic" geometry. In fact, AutoCAD could not handle the "warped" geometry used in Rhino. In other words, the latest version of the building was not just designed in AutoCAD, it was designed *for* AutoCAD.

Thorndike's decision to turn to AutoCAD was as much about preserving his creative role on the project as it was about making a design that conformed to the firm's ideals and was a rational use of resources for the client. He could have asked another architect to do the work that he could not do with Rhino, but unlike Shales and Morris, Thorndike is carving out his identity in the firm by learning new computer applications. He says that asking others to make a geometric model for you takes longer. He does not directly say that it makes you dependent on them, but this thought is present in everything he says. "First of all, you have to wait for that person to become available. Then you have to explain the whole project to them. Lastly, if you want to hold their interest, you have to give them something significant to work on. I would rather do it myself." In the course of working on this project, Thorndike improves his ability to move back and forth between warped geometries in Rhino and platonic geometries in AutoCAD. After one weekend he says, playfully using childhood grammar, "Me and technology worked together this weekend."

Thorndike feels strongly that learning Rhino will help him extend his role in developing competition work with Morris. However, he has worked in the office for twelve years and knows that within the Paul Morris office, his role in design will extend only so far. Eventually, he wants to venture out on his own. For Thorndike, working at Paul Morris is an opportunity not only to learn about design but about the mechanics of owning a firm and managing projects. For Thorndike, skill at geometric modeling enables him to negotiate a role in the office, a "chip" that provides access to certain tasks. However, Thorndike stresses that for his long-term development, other skills, such as project development, may be equally or more important for him. He must continually negotiate his roles as designer, simulations expert, and manager in order to assure a creative role for himself in contemporary practice.

RALPH JEROME ARCHITECTS

Ralph Jerome Architects, a hundred-person office in the Midwest, uses CATIA to realize complex projects that require it to manage the knowledge of many partnering disciplines. Here, CATIA serves as a "place" where many different kinds of knowledge meet. The details of construction are often contributed by consultants and fabricators. CATIA brings together knowledge of construction materials from outside collaborators with schematic information about designs. Ralph Jerome stresses that, for him, this technology brings the architect closer to the craftsperson who will actually handle the materials of construction. He sees CATIA as a way to bypass the many layers between sketch and final building. However, at Ralph Jerome Architects the reconfiguration of work around CATIA has also led to confusion, redundancy, and loss of data, problems that are common when working with digital files. These developments have prompted the establishment of a new role. Like Paul Morris, Ralph Jerome needs a "keeper of the geometry," someone to play a role similar to, but more powerful than that of Robert

Laird. At Ralph Jerome's firm, this role has a formal title, "Director of Computing." Malcolm Dietrich has this role and is responsible for coordinating the geometric modeling work in the office.

For Dietrich, the computer is on its way to becoming the unifying collaborative space for designers and technicians. In Dietrich's vision, one person is at the center of the building process whose power comes from access to knowledge, all of which is in the computer. Dietrich describes this person as a "master techie-enabled architect sitting in the middle."

As Dietrich envisages it, the techie-enabled architect has the potential to crystallize a new kind of integration among members of the firm and external contractors. The techie-enabled architect understands every bolt in the building and can see and coordinate the work of every person involved. Dietrich looks forward to the day when he will be able to say, "I understand what this shape is and how it's built and how pieces go together, and I can validate and I can stand up and say this thing will work." On one level, the architect at the center is all-powerful; yet there is a paradox in Dietrich's vision. In this model, craftsmen are given a more creative role in design and sometimes Dietrich talks about himself as "just stitching the bits together." Being in control of the machine can seem like low-level work, "Sometimes I feel I'm just a full-time translator."

TECHNOLOGICAL ROLES: FROM TRADING ZONES TO COMMUNICATIONS MEDIA

In the stories of these architects we see a variety of responses to technology and a variety of emerging identities. Quix is a new breed of expert technician; Shales defines herself by her refusal to participate in the demands of technology; Morris and Thorndike are forging a new dynamic for master and apprentice in a culture of simulation; Dietrich provides a glimpse of the architect on the verge of becoming cyborg, an identity that requires becoming one with the machine.

There are elements of Dietrich's vision that already are at work at Ralph Jerome Architects. For example, Dimitri Kabel, an architect at the firm, "captures" in CATIA the specific physical knowledge that fabricators bring to the design process. For Kabel, working with computer simulation has made him see the whole of architecture as being about simulation, something that was not clear to him when representations were done with pencil and paper. When a project needs special technical knowledge about a suspended glass wall, Kabel works with a curtain wall consultant to model this knowledge in CATIA; the geometric model serves as a communications medium within the office. Yet, when the consultant is asked, for example, to codify his conventions for working with glass, the current limitations of Dietrich's vision become clear. The process of fabricating this curtain wall cannot yet be described in a form that can be modeled in CATIA. Dietrich laments that "one of the hardest things is going to a fabricator and telling them, provide us with your rules; what are the rules that we need to adhere to [to] make this thing buildable?"

In these stories we see that the question, "Who am I in relation to software?" has become central to how architects negotiate professional identity. Paul Morris and Ralph Jerome use it as a tool to make buildings in partnership with computationally savvy colleagues. But some architects are primarily masters of the virtual, securing international reputations in architectural competitions through objects that exist only on the screen. Their practices flourish on computers and in art galleries rather than on construction sites. Among them are architects who work with master programmers who are sometimes also architects. Such is the relationship between Tom Haig and Mike Orlov, both of whom work in an academic setting. Haig understands how the program operates and has a working model of its underlying operation but cannot program the simulations that make his work possible. This role falls to Orlov, Haig's master programmer. Haig is master of a realm he cannot completely enter. Orlov says of their partnership: "It's really funny, while we are communicating together about the same thing, we are talking about different levels of the same thing."

Here, as in other relationships among architects who approach geometric modeling with different levels of understanding, the technology acts as what historian of science Peter Galison has termed a *trading zone*. Galison uses the term to describe how those who belong to different social groups can productively trade objects and information without having the same understanding of the exchange.[13] But it is an apt description of how architects work across their own cultural divides whether it is Morris, the master, and Thorndike, the apprentice, or Haig, the virtual architect, and Orlov, his hacker partner.

Galison's terminology stresses the diversity in what different architects bring to negotiations with technology; another way to look at their exchanges is to stress similarities in what the software imposes on all of them. A program such as CATIA embodies culturally specific views of image making. These will be transmitted to any user who engages with the program; that user will be drawn into the embrace not just of a specific program but of a way of encoding reality.[14] Architectural theorist William J. Mitchell talks of software as "frozen ideology."[15] Sociologist Gary Downey describes the power of computer-aided design as the subordination of design to computer graphics, a process in which one's eyes and fingers are given over to the machine.[16] In Michel Foucault's language, the program becomes a new kind of power, a force that "produces reality; it produces domains of objects and rituals of truth."[17] This is what Quix recognizes when he speaks of CATIA as more than a program but as a way of seeing the world.

Like any profession, architecture may be seen as a system in flux.[18] However, with their new roles and relationships, architects are learning that the fight for professional jurisdiction is increasingly for jurisdiction over simulation. Computer-aided design is changing professional patterns of production in architecture, the very way in which professionals compete with each other by making new claims to knowledge. Even today, employees at Paul Morris squabble about the role that simulation software should play in the office. Among other things, they fight about the role it should play in promotion and firm hierarchy. They bicker about the selection of new simulation software, knowing that choosing software implies greater power for those who are expert in it.

As we have seen, sharing a screen does not necessarily mean sharing a vision. It does bring, however, a new kind of intimacy that makes explicit what is shared and what is not. For even when software is celebrated and used creatively, certain former habits of mind endure. In particular, architects remain preoccupied with drawing, the expression of another kind of intimacy with volume and profile.

Thorndike, the apprentice, believes Paul Morris hired him not for his technical expertise but for his ability to draw with a pencil.

For Morris himself, who thinks with a pencil and is newly bound to an apprentice who can help him think on the screen, there are already intimations of a convergence. He has, after all, chosen a working partner who is both fluent on the screen and with a pencil. In an instrumental sense, it is Thorndike's ability with a computer program that makes him indispensable to Morris, but it is Thorndike's fluency with a pencil that probably makes Morris comfortable with him as a colleague.

Quix, the CATIA expert, has had the opportunity to work with both firms discussed in this essay. He sees Paul Morris and Ralph Jerome as accepting new technologies in different ways, both for their firms and for themselves. Morris is more personally receptive to the computer. When it comes to the machine, Quix says that "[Paul Morris] is better at looking." In contrast, Ralph Jerome, keenly aware and a bit wary of the power of the technology on which his office is making its reputation, refuses to look at his designs on the screen. He says it is like "putting his hand in the flame." Morris, Jerome, and those who work with them are in a continual struggle to define the creative roles that can bring them professional acceptance and greater control over design. New technologies for computer-aided design do not change this reality, they become players in it.

PERFORMING THE PROTEIN FOLD

Natasha Myers

Scientists rely on models—from physical constructions to virtual animations—as objects-to-think-with.[1] Changes in the cultures of modeling reflect profound changes in research fields. Models show science in action, with all of its reliance on intuition and serendipity, with all of its engagement of the scientist's body as well as mind. In research and pedagogy, scientists do not simply refer to models, they *perform* models.[2]

Historians of science David Kaiser and Maria Trumpler found that textbooks document and illuminate the history of changing scientific models and pedagogical styles.[3] Here, an ethnography of a biology class on protein folding enables me to focus on a more ephemeral element of pedagogy—the role of models in the dynamic "performativity" of classroom teaching. Invoking "performance" rather than "representation," I am able to explore the ways that multidimensional models come to dwell in scientists' imaginations.

My story begins with new developments in modeling practices among molecular biologists. These days, after decades of research "decoding" the linear DNA molecule, the object of inquiry is shifting. Multidisciplinary teams of scientists and engineers now converge on different objects: protein molecules. A new generation of life scientists must learn to visualize these complex, changing, and minute molecular substances. This requires new cultures of pedagogy and training to shape a new generation of protein scientists.[4] One of the key skills novice researchers must acquire is a "feeling" for how protein molecules move and interact.[5] This is a feeling that experienced researchers need to pass down to students. To do this, teachers encourage their students to engage models actively and physically. In this way, teaching is a practice of informing students' minds and forming their bodies.

MODELS, BODIES, AND IMAGINATIONS

Scientific representations are important, but equally so are the ways that images and models come to dwell in scientists' imaginations. When Trumpler studies the development of rich imaginary landscapes among molecular biologists, she focuses on their changing representations of membrane channel proteins, structures that have been key to molecular neuroscience over several decades. Trumpler writes about the "importance of the scientists' own mental images," and how they "privately conceived" of "large, complex, 3D molecule[s] moving in time and space."[6] For Trumpler, multiple "complex mental images" of otherwise invisible substances help researchers pose experimental questions and communicate with each other. No single model could suffice.

Trumpler asserts that the successful mental image of proteins is dynamic, generated by what she calls a "convergence" of many distinct modes of representation.[7] Mental images are thus the product of different kinds of representations that shift as new visualization techniques and conventions are introduced. How to access the history of this rich visual imagery? Trumpler uses the diagrams and models in textbooks because they tend to combine distinct visual representations in a single figure. For Trumpler these figures are templates that convey the mental images expert scientists wish to elicit in the imaginations of their students.

Trumpler makes significant contributions to our understanding of the nature of scientific imagination. She identifies the kinds of sources historians might use to find expressions of otherwise imagined entities; she shows how scientific imaginations are produced through pedagogical processes. Here I explore how moving from history to ethnography and moving from the study of two-dimensional representations in textbooks to 3D models in classrooms can extend and deepen our understanding.

With Trumpler, I propose that 3D models have both a rich material history, as well as a lively inner life. But her use of the term "mental image" does not fully convey the degree to which scientific imaginations are involved with the body and it does not fully convey their multidimensional nature. To understand scientific imaginations, one must go beyond what scientists write or put in textbook illustrations. Models are not just representations or results; they are "enactments" and they are "built to be engaged, inhabited, [and] lived."[8] Scientists do not just hold their models in their minds; they carry their models within their bodies. The engaging and laborious experience of building and working with models is the means through which researchers *incorporate* knowledge of the form and structure of their models.[9] In this process, scientists' bodies become instruments for learning and communicating their knowledge to others.

To study this process, I turn to sites where scientists endeavor to communicate embodied molecular knowledge. One of these is a semester-long lecture course that examines the biology of protein structure and folding taught at an East Coast research university. The professors for the course are Jim Brady and Geoff Miller.[10] Brady is a prominent protein scientist who is also known for his commitment to science education. Miller is a mechanical engineer who has recently taken an interest in protein structures. Because there is no unified model to describe how proteins fold, a course on the subject offers the perfect site to study how multiple representational forms converge.

In the first weeks of the course, students are encouraged to learn the intricate molecular structures of each protein "by heart." Brady tells the class: "We want you to have it in your head. You need to know it cold." Through commanding performances, Brady and Miller work hard to impart the skills students need to get protein structures in their "heads" and "hearts." While Brady spends much time rapidly scrawling biochemical equations and experimental data across the board, he also experiments with analogies and demonstrates protein structures with colorful ball-and-stick models. Miller brings his engineering expertise to the classroom, using dazzling interactive computer graphics displays to demonstrate the special features of each fold. And beyond all of this, the two professors pull 2D representations off the page and into real-time performances of molecular form. In this way, they show their students how to use their bodies as a resource in order to understand complex 3D molecular structures and how to build their own embodied models, layer by layer.

THE PROTEIN FOLDING PROBLEM

The "Protein Folding Problem" is cross-listed in the departments of biology and chemistry, and the students, both undergraduate and graduate, come to it from diverse disciplinary backgrounds including physics and computer science. Over fifty students show up to its first meeting. Yet just fifteen years ago, only ten students took the class. Brady remarks on this to the class and asks, "What has changed? Why so much interest in proteins?" Brady believes that proteins have captured scientists' imaginations because these molecules account for the "second half" of the genome. Also, interest in proteins is spurred by the belief that studying them will have practical consequences, for example in the production of the protein insulin, as well as drugs to treat Alzheimer's, Huntington's, and CJD, the human form of "mad cow" disease. All of these are protein folding diseases. In each of them, misfolded proteins damage tissues by producing lethal aggregations inside cells. Medical progress thus depends on research in this area.

In class Brady describes the history of how scientists learned about protein folding. The defining moment for him was in 1957 when the first model of a protein was produced by John Kendrew's laboratory at the Laboratory of Molecular Biology at Cambridge University. Using low-resolution X-ray diffraction data, a 3D model of myoglobin (the protein that carries oxygen to muscle) was built from ready-to-hand materials. Nicknamed the "sausage model," its winding polypeptide chain was crafted in Plasticine and supported on thin wooden sticks. According to Brady, this model was "shocking" to the scientific community: "The moment that image hit the press people wondered, how does this chain know where to go?" Kendrew himself was surprised by the irregularity of the protein structure that emerged from his X-ray data. He had expected that proteins, able to organize themselves into highly ordered crystalline arrays, would have some form of internal symmetry.[11] What gave this strange little protein its conformation, its special fold?

Brady describes the complexities of the protein fold: "You can have a deep understanding of the end state, but have no clue how it got there." Protein folding occurs at the scale of atoms and the speed of nanoseconds. Visualizing it presents a challenge to researchers and they have responded with an array of imaging, modeling and simulation techniques drawn from chemistry, physics, crystallography, molecular genetics, mathematics, and computer science. It has, however, resisted definitive visualization. No single representation captures the elusive behaviors of a polypeptide chain as it wriggles its way through a range of conformations in search of its "active" or "native" state.

The process of protein folding is difficult to communicate through pictures and words and also presents a challenge to the ubiquitous application of cybernetic models of communication to cellular processes.[12] If one follows the mechanism outlined by the increasingly shaky "central dogma" of molecular biology, polypeptide chains are the end product of a complex process that involves the "transcription" and "translation" of "information" stored within the genome.[13] In this model, RNA serves as a "messenger" to transcribe the message contained within the nuclear DNA. Once transported into the cytoplasm, ribosomes, tiny macromolecular organelles that "read" the ribonucleic transcript, "translate" it into a long polypeptide chain of amino acids linked end to end.

It is at the point where the polypeptide chain is released into the cytoplasm that protein folding emerges as a practical problem for the cell. This is also the point where the metaphor of cybernetic communication breaks down. For a protein to acquire cellular activity, the long, floppy polypeptide chain must be folded into a complex, 3D structure. There is no template or code that determines a protein's active form: many different amino acid sequences can produce similar tertiary structures, and similar sequences can produce different folds. Immersed in the watery and chemically active environments of the cell, proteins *figure out* how to fold themselves into their active 3D forms. It is this interruption of the cybernetic model of seamless information flow that has stumped protein folding researchers for the last fifty years.

Since the early 1960s researchers have searched in vain to determine how amino acid sequences can be used to predict the protein folding pathway and the active structure of the protein. Biologist Cyrus Levinthal, working at MIT between 1963 and 1967 in collaboration with MIT computer scientists and engineers, was the first to devise an ambitious project to use computer algorithms to solve complex protein structures.[14] Working with the assumption that the biologically active conformation of the protein would be at its lowest energy state, Levinthal developed computer algorithms that defined the shapes of molecules.[15] However, no matter how much he tweaked his computer programs, he failed to predict the correct conformations for known protein structures. Increasingly powerful computers are currently being applied to protein prediction.

Brady, however, is skeptical of those who believe they would find the answer to the protein folding problem with bigger and better computers. As he explains to the class on the second day, DNA sequences cannot be treated as a proper language, and so, the rhetoric of code and information does not hold for proteins. Brady resists the idea that researchers can conjure algorithms to predict the unique fold of each protein. If there is no algorithm to predict the fold, what then could be going on in the transformation of the linear peptide into the complex 3D structure of the mature protein? For Brady, protein folding is a "deep problem." Refusing to reduce protein folding to a two-dimensional problem of reading or writing DNA sequences, he draws his students' attention to the depths and dynamics of the fold. Investigations into protein folding demand active bodies and imaginations, and Brady takes seriously his pedagogical role: he must teach a new cohort of scientists the skills to use their embodied imaginations to help solve the protein folding problem.

ENFOLDING THE FOLD

Joanna Bryson worked with Brady as a graduate student, and identifies as a "true protein folder." Currently she works as a postdoc to develop new strategies for teaching biology to undergraduate students. When she was an undergraduate she studied chemistry, and early on in her training realized that she had a gift for visualizing 3D molecular structures and chemical interactions. She credits her success in chemistry to her skill as "visual spatial learner." "I was never a memorizer. . . . But from the moment that they put Van der Waal radii on molecules . . . I could see it in my head"[16]:

You know, when two molecules come together, or even unlike molecules that don't want to be next to each other, for whatever reason, I can see those electrons moving to the other side of the molecule. It made total visual sense to me. It didn't make sense on paper. It never made sense on paper. I wasn't a memorization reaction learner at all. But electron-pushing diagrams made so much sense. And I kept wondering, "Why is everyone else having such a hard time with this?"

Bryson laughs when I ask her to describe what these models look like in her imagination:

NM: Are they colored?
JB: Yeah! [laughter] I've never thought about it this closely. Yeah, they're colored! That's kind of bizarre.
NM: Are they textured?
JB: No, they are pretty smooth. Yeah, they're pretty smooth! [laughter] My proteins are either ribbon diagrams in my head or Van der Waals, depending on what I'm looking at. If I'm looking at a binding pocket, it'll be a Van der Waal image with just a surface that I see up there doing it. If it's a folding, if anything has to do with folding, it's a ribbon structure. It's always a ribbon if it's folding.

Bryson recognizes that the models "in her head" conform to conventional representations that she has learned. Diagrams that visualize Van der Waal radii are a particularly interesting convention: they not only describe the volume that an atom will occupy, but they are flexible spheres that can also be used to model regions of attractive and repulsive forces between atoms. Some researchers say that they experience Van der Waal forces viscerally: when they look at structures whose atoms defy allowable radii, they feel the intermolecular tension in their bodies.[17] Bryson is one of these people. "It's hard to describe," she tells me, "I just see it in 3D." For Bryson, "seeing" is a corporeal experience. As she describes proteins that she has worked on, her body comes alive with gestures that demonstrate molecular forms and forces with which she is intimately familiar.

Bryson describes how she communicates the complex molecular forms and movements that inhabit her imagination. She begins by holding her arms out in front of her, her wrists touching, palms open and facing upwards. Her hands seem to hold an invisible substance. Then she says:

I always do this [emphasizing the gesture]. Whenever I talk about the crystal structure . . . I always do this. 'Cause that's how the molecule kind of looks. It's like this [emphasizing the gesture, she rotates her hands and body]. You know. And domain one unfolds, and then it's flopping around. [She mimes this floppy domain with one hand.] You know. Always. Always. Even in my thesis defense talk. It was like this. And it was flopping around like this.

In attempts to communicate protein folding to her colleagues, Bryson performs her embodied model. However, as a teacher trying to get her students to learn molecular structures, Bryson realizes that not everybody has this ability to visualize complex objects in three dimensions:

To me it is so intuitive. My hardest stumbling block is to rationalize that it's not like that for everyone else. So to me the hard part was actually stepping back, and realizing not everyone gets this. How can I get other people to understand what I see in my head automatically?

She recognizes that 2D textbook images are not enough for her students to learn molecular forms. Professor Brady is faced with a similar problem. He wants students to have instant recall of molecular structures: "You must be able to see leucine [an amino acid] in three dimensions. You need to be able to see it immediately." When he demonstrates the structure of the leucine zipper, an important fold for protein-DNA interactions, Brady jokes that students need to know it well enough that "if your grandmother asks you for it, you could draw it": "We want you to have it in your head. You need to know it cold."

How then does 3D knowledge of protein structures get into students "heads"? Brady's animated teaching during his protein folding class offers some clues. What Brady and Bryson demonstrate is that there is nothing "automatic" about the process.

In one of his early lectures Brady draws students' attention to some confusion around a homework assignment. Some students had trouble with the wording of a question. Directing the students to a ribbon diagram in their textbook, Brady asked the students to "draw, copy, or trace a version of figure 2(e) with the alpha carbons and nitrogen atoms clearly labeled or colored."

Some students had trouble interpreting the meaning of "copy." Brady clarifies: "This means *hand copy*! If you [photocopy] it, you don't assimilate it!" He demonstrates for the class. Against the background of a larger-than-life projected ribbon diagram of the protein structure from the textbook, Brady's entire body is swept up in the act of tracing the elaborate curvature of the fold. He tells the class: "You have to signal actively" in order to "get" the subtleties of the fold. He effectively shows that "you can't not learn something" if you get your body "actively" involved.

Bryson has learned a lot from Brady's approach to teaching. She and her colleagues in the biology department have recently developed a series of workshops and lesson plans using specially designed 3D models and interactive computer graphics to help students learn to visualize the structures and movements of biological molecules. Through this work, Bryson has come to recognize that teaching these concepts places extra demands on her body to perform the multidimensionality of biological phenomena. Bryson is wary that students will take her body-folding performances too literally (she calls this "anthropomorphizing the molecule"). Nevertheless, she animates molecules with her body in class:

I probably like the dancing and movement so much [in the classroom] because I do see these things rolling around in 3D in my head. And yeah, it's like, if I could get my body to do this [as she curves her body around an imaginary fold, she also voices the movement: "Shwooo!"], and have this little arm flapping in the breeze. I don't know. It just makes more sense.

As she talks, her body comes to life, and I can see her delight in communicating the details of the fold. Brady and Bryson are not alone. In both professional and pedagogical contexts, scientists often perform the structures and movements of proteins through gestures that use their entire bodies—hands, arms, shoulders, torsos, and even legs.

MODELING BODIES BY ANALOGY

During my semester in the protein folding class I saw lessons on the biochemistry of protein folding illustrated by graphs, equations, animations, and models both physical and virtual. Brady augmented these with lively analogies that drew students into new kinds of understanding.

Some theorists suggest that modeling by analogy and metaphor enable "associations from one field to animate a scientist's thinking about another field," and so make "research 'do-able.'"[18] Analogies can be thought of as producing their own form of "realism."[19] Brady uses analogies in his class to make protein folding more accessible.

Brady offers students what he calls a "motivator" to capture their interest in the protein of the day and to help students get a better feel for these substances. For example, Brady talks about cooking egg whites and making Jell-O to illustrate effects of heat on proteins: heat denatures proteins and promotes their aggregation, changing the state of a substance from fluid to solid. To call up people's familiar experiences with collagen in its denatured form, Brady offers other examples such as glue, the congealing of chicken soup and the healing of wounds. He also uses fever as an analogy to illustrate the temperature sensitivity of protein folding ("above 106 degrees, you're dead"). He employs human-scale analogies like hair, gastric juices, and the curdling of milk to model the structures and states of proteins at the molecular scale. Analogies that take students from the macroscopic to the molecular are *metonymic* in their nature, connecting part to whole. However, rather than using the part to stand in for the whole, Brady uses macroscopic phenomena to draw students into microscopic worlds. In this way he gives texture to the molecular worlds that the students in the class might not otherwise grasp.

Brady recognizes that his "motivators" are not compelling for every student. He uses leather as an analogy to give his students a sense of the texture of collagen in its denatured form. When he doesn't get a rise out of the class, he mutters, "Nobody wears leather shoes anymore. We're moving into the modern age. I need to change the motivator!" And again, when no one can answer his question about what happens to blood when you heat it up, he feels compelled to provide washing instructions to the students so they know what to do if they get bloodstains on their clothes:

Let's say we want to look at denaturation versus temperature. Let's take a mixture of ovalbumin and lysozyme in egg whites. What happens when you heat that up? Does it denature? Yes. Can you establish equilibrium between the denatured state and the native state? No? What about hemoglobin? Some of you are very familiar with hemoglobin. What happens when you heat up hemoglobin? Some of you must have had bloodstains? You never had bloodstains? Wow. [A male student tentatively answers: "It loses color?"] It changes its color. Right? And then is it easy to get out? You know? No! Once you heat up hemoglobin, right . . . if you make a mistake and you've got a bloodstain, and you put it in a washing machine with hot water, you are done, you'll never get it out. You are much better off washing it in cold water [laughter]. Let me tell you, you ever get a bloodstain, you have to wash it with cold water! [Brady gets louder.] If you wash it in hot water you will get thermal denaturation and aggregation!

This is, of course, a lesson that the women, who make up about half the class, likely understood quite well, although they were reticent to admit it publicly. Faced with an awkward silence from his students, Brady began to search for hooks that would elicit recognition and produce an embodied understanding of the effects of heat on protein folding. When the hooks work, the effect can be visceral. I experienced this when contemplating my own scars, imagining how collagen fibers form at the site of fresh wounds. In that moment my attention was drawn to my knee, and my memory of watching a doctor stitch it up after a childhood accident. Shifting between scales, I could remember the sting of the stitches, and simultaneously visualize and feel minute collagen threads rapidly extending and coiling up around each other, closing the wound and binding the tissues tight. Effective models, like analogies, can produce what philosopher of science Isabelle Stengers calls "lures," that is, abstractions that "vectorize concrete experience."[20]

THE BODYWORK OF MODELING

Molecular models embody their makers' intimate knowledge of molecular form. When molecular models leave the hands of their makers, they become teachers in their own right by attracting the active and "curious hands" of their users.[21] During a lecture on secondary structures in proteins, Brady holds a ball-and-stick model of an alpha-helix. Well worn and well loved, this relic was brought to this campus forty years ago from Cambridge, England. It is a model of the molecular structure that Linus Pauling discovered in 1948. As the story goes, Pauling came up with the structure by folding and refolding paper chains of amino acids while lying in bed recovering from the flu.[22] To build the model, Pauling drew on his knowledge of chemical laws and the constraints of how amino acids fit together. But in the process of discovery, he played with ready-to-hand materials and figured out the structure of the alpha-helix in an improvisation that engaged both his body and imagination.

Now, in Brady's hands, this replica of Pauling's model takes on a new life; it is at a different stage of its pedagogical life cycle. Brady tells a story about his experience with this model when he was a post-doctoral fellow. He worked at the Medical Research Council in Cambridge, England, at the time when Cyrus Chothia and Arthur Lesk were investigating Pauling's structure. In this context Pauling's model was transformed. No longer an object whose intrinsic properties were to be determined, an "epistemic thing" in Hans Jörg Rheinberger's terminology, the model became a "technical object."[23]

Made over into a tool, it became a means to produce new kinds of questions and insights.

Brady and his colleagues built physical models of the alpha-helix and tried to figure out how the helices interacted. They spent "years and years and years just *looking* at this structure." Brady says:

One thing was very clear in that group: some people were just able to sit and look at the structures. But most people could not do that. They had to get up and get a cup of coffee and do an experiment. Some people could just look at the structures. And finally they saw things that nobody else saw. Because that discipline of sitting and looking is something that is very hard. And it is something that has been lost. I worked with Aaron Klug who won the Nobel Prize for a three-dimensional structure. He used to sit there and say to us, . . . "You Americans you can't sit still long enough! You go off and do an experiment. . . . You don't *look.*"

When Brady shows Pauling's model to the class he makes it clear that "sitting and looking" at a model is not a leisurely activity. You can't just gaze lazily at the model: "If you *just look* at it you don't see anything." Playing on the double meaning of the verb "to grasp," he tells the class: "Now . . . this is not easy to grasp, and that's why it's so important to *grasp* these structures. He picks up the model, which is two feet tall and a foot and a half in diameter, and rotates it around in his hands, telling the class, "soon you will start to see." With his hands and eyes he shows the class how to "walk through it" amino acid by amino acid. He explores the model atom by atom and feels out the grooves and ridges formed by the side chains as they spiral up the helix.

Brady demonstrates that "sitting and looking" involves the body. Teaching protein folding becomes, then, a practice of "articulating" students' bodies and imaginations with intricate knowledge of molecular form.[24] Brady's class is a rite of passage into an emerging professional identity that demands new corporeal practices.

MOLECULAR GESTURES

Psychologist and linguist Elinor Ochs and her colleagues study gesture in the performance of scientific concepts, describing the gestures that mediate communication among physicists as they convey their research to each other in weekly lab meetings. The Ochs team is interested in tracking "understanding-in-progress," and so it observes how bodily gestures help physicists narrate and dramatize their scientific stories. Ochs reveals physicists' gestures as a "dynamic grammar," a means of supporting language, which helps them to make statements about mathematical relations and 2D graphic displays.[25] In contrast, Brady's gestures perform objects that take up space and move through time—objects that defy simple description. Brady encounters the limits of language when he tries to find words to paint a portrait of dynamic, 3D molecules:

It is clear from the X-ray diffraction patterns that proteins are objects with space in them. This is very different from packed polymers. So, we can ask: What is the character of the interior? Is it oily? Is it patchy with regions of solvent? But patchy is a two-dimensional word. I can't think of a three-dimensional word that gets at this.

As we have seen, where language and 2D images fail, Brady uses his body. Curving over and tucking inward to create a concave form, he uses the shape of his arms to mimic the internal organization of helices and sheets. When describing the packing of two helices in a protein, he repeatedly draws his arms in towards each other, crossing them at the forearms to specify the precise angle at which they are associated. The flexibility or inflexibility of this association is made clear through the tension he holds in his muscles.

Brady's gestures convey the form and movements of the molecule through the form and movements of his body. As such, he performs a kind of *mimetic modeling*. In this sense Brady *becomes molecular*,[26] a productive form of anthropomorphism.

Mimetic modeling is a precise practice. During a demonstration of the packing of helices during protein folding, Brady holds his arms out in front of him, crossed at the forearms to mimic how it is that the "side chains are talking to each other": "Now when two helices are packing against each other they form a junction . . ." He pauses, looks up, and gestures to a male student in the front row. "I want you to stand up." "All right," the student says and rises. "Now, point to the junction," he says. The student points vaguely at Brady's crossed arms. "No, not there!" Brady uses his voice and eyes to redirect the student. "Right between . . . Yeah, okay." Brady carries on with his description while the student stands by his side, pointing at the junction. In this moment, Brady makes explicit that *he is* the model.

The protein folding course was, then, not so much designed to transmit a set of known facts, but to equip students with the practices they will need to push the field forward. The course gives students a mission (Brady says, "Hopefully one of you will solve this class of problems") and a method (adding, "first you must master the structure, then you can move on to experiments"). For this kind of mastery, Brady's students must to be willing to let molecular models instruct their bodies so that they can embody the fold.

NOTES

OUTER SPACE AND UNDERSEA

William J. Clancey, Becoming a Rover

1. The membership of the MER science team is public; the conventions in this volume are to provide pseudonyms. Ages are at the time of *Spirit's* landing on Mars, January 4, 2004.

I am indebted to the six MER scientists who shared their personal experiences in interviews and reviewed the text of this chapter. Oscar Biltmore, Bettye Woodruff, and Karl Trainor also helped orient me during the nominal mission at JPL in January-February 2004. The MER Human-Centered Computing ethnography team that I advised at NASA/Ames included Charlotte Linde, Zara Mirmalek (University of California, San Diego), Chin Seah, Valerie Shalin (Wright State University) and Roxana Wales; their observations and our conversations played a crucial role in my understanding of MER operations. This work has been supported in part by NASA's Computing, Communications, and Information Technology Program, Intelligent Systems subprogram.

2. The initial planned operational period of a mission is known as the "nominal mission." At sol 25, *Opportunity* is still in this phase and will be so for three months. During the nominal mission, its scientific team lives in Pasadena. In the six months after the nominal mission, the team works closely together, joined electronically, from their home institutions. In the years that follow, a rover team becomes fully distributed. My report is based on observations of MER science teams in 2004 and follow-up interviews in late summer 2006 with six MER scientists from diverse aspects of the program and at different points in their careers. Broadly speaking, the MER scientists are planetary scientists focused on the exploration of the solar system. But each gives a more specific answer when asked "What kind of scientist are you?" In profession they range from astrogeologists to specialists in artificial intelligence and robotics to planetary geologists to biogeochemists.

3. Donald Schön, *Educating the Reflexive Practitioner* (San Francisco: Jossey-Bass Publishers, 1987).

4. Sherry Turkle, "Whither Psychoanalysis in Computer Culture," *Psychoanalytic Psychology: Journal of the Division of Psychoanalysis, American Psychological Association* 21, no. 1 (Winter 2004): 16–30.

5. Steve Squyres, *Roving Mars: Spirit, Opportunity, and the Exploration of the Red Planet* (New York: Hyperion, 2005), 328, 334, 336.

6. Ibid., 81.

7. This way of proceeding stands in contrast to how telescopes such as the Hubble are managed. There, proposals by investigators and small teams requesting particular observations are made months in advance. Here, with the MER, the scientists move together in their exploration of Mars, making decisions each day and usually getting the results within a few sols.

8. Simulation plays a central role in the space program, ranging from computer models of spacecraft trajectories to full-scale physical mockups of cabins and modules to chambers and devices that replicate the radiation, cold, microgravity, and near vacuum of space. Simulation for the MER scientists

included multiple-day simulated missions for training and refining operations, as well as rover "test beds" for practicing and testing MER behaviors. In focusing on the scientists' personal experience, this chapter considers just one small part of the nature and importance of simulation throughout the MER mission's planning and operations.

9. In *Educating the Reflexive Practitioner*, Donald Schön emphasized how designers effectively coordinate developing concepts with physical artifacts and models through a reflective and manipulatively iterative process of "seeing as" and "conversation with materials."

10. Squyres, *Roving Mars*, 325.

11. Invited address, Mars Society Annual Convention, Washington, D.C., August 2006.

12. JPL *Spirit* Update August 25, 2006, available at <http://marsrovers.jpl .nasa.gov/mission/status.html> (accessed October 21, 2008).

13. Three and half years into the journey, the "chemists" versus "geologists" conflict has softened significantly, although Biltmore still wants to drive and drive!

14. Within the planetary science community, scientists are specifically those who work in the fields of inquiry that drive space exploration, especially geology, physics, chemistry, biology, and astronomy. Psychologists, ergonomists, and social scientists are often called human factors specialists. Thus scientific work outside of planetary science is viewed only instrumentally. For MER, as for all planetary science missions, *scientists* are defined as those people who know how to gather and interpret the instruments' data; their expertise relates to the scientific purpose of the technology. Correspondingly, *engineers* are defined as those who know how to make, test, package, and control the instruments; their expertise relates to the manufacturing and operation of the technology. See W. J. Clancey, "Field Science Ethnography: Methods for Systematic Observation on an Expedition," *Field Methods* 13, no. 3 (August 2001): 223–243.

15. Conflicts have been known to occur between scientists and engineers on oceanographic expeditions. H. Russell Bernard and Peter D. Killworth, "Scientists and Crew: A Case Study in Communications at Sea," *Maritime Studies and Management* 2 (1974): 112–25.

16. Squyres, *Roving Mars*, 246.

17. Ibid., 251.

18. Ibid., 377.

19. Ibid., 377.

Stefan Helmreich, Intimate Sensing

1. Sigmund Freud, "Civilization and Its Discontents," in *The Standard Edition of the Complete Psychological Works of Sigmund Freud*, trans. and ed. by James Strachey, et al. (London: The Hogarth Press and the Institute of Psychoanalysis, 1953–1974), vol. XXI; Jacques Cousteau with Frédéric Dumas, *The Silent World* (New York: Harper and Brothers, 1953).

2. Charles Goodwin, "Seeing in Depth," *Social Studies of Science* 25, no. 2 (1995): 237–274. Drawing on 1990s fieldwork conducted on an American research vessel off the Brazilian coast, Goodwin describes how scientists from diverse disciplines collaborate with one another and with a ship's crew to fashion portraits of the sea beneath. His narrative has at its center a commonplace piece of oceanographic equipment called a CTD, a package of sensors that generates readings of ocean conductivity, temperature, and depth (hence "CTD") and that also takes samples of seawater as it is lowered by pulley from the side of a research ship into the water column. In onboard control rooms, scientists monitor scrolling computer screens of data sent from the CTD as the device travels through different depths, a descent they guide in coordination with the ship's crew, who work on deck, operating a winch. The whole system affords scientists "perceptual access to the world they are sampling, while simultaneously shaping what they are able to see there," 250.

3. Hillel Schwartz suggested this term to me.

4. Sabine Höhler, "Floating Pieces, Deep Sea, Full Measure: Spatial Relations in Oceanography as a 'Field Science.'" Paper presented at the meetings of the Society for the Social Study of Science, Cambridge, Mass., November 2001.

5. Sherry Turkle, Joseph Dumit, David Mindell, Hugh Gusterson, Susan Silbey, Yanni A. Loukissas, and Natasha Myers, "Information Technologies and Professional Identity: A Comparative Study of the Effects of Virtuality," in *A Report to the National Science Foundation on Grant No. 0220347* (Cambridge, Mass.: Massachusetts Institute of Technology, 2005), 2.

6. David Mindell, "Between Human and Machine," *Technology Review* (February 2005), available at <www.technologyreview.com/computing/14171/> (accessed October 22, 2008). I thank David Mindell for commenting on an earlier draft of this essay.

7. The membership of MBARI science teams and ship crew is public. The convention of this volume is to provide pseudonyms.

8. See <www.mbari.org/cruises/lobos/map_image.html> (accessed October 22, 2008).

9. James Hamilton-Paterson, *The Great Deep: The Sea and Its Thresholds* (New York: Random House, 1992), 168.

10. Goodwin, "Seeing in Depth," 239.

11. Alain Corbin, *The Lure of the Sea: The Discovery of the Seaside in the Western World 1750–1840.* Translated from the French by Jocelyn Phelps. (Berkeley: University of California Press, 1994; French edition, 1988).

12. Goodwin, "Seeing in Depth," 256. Classical sociologies of relations between scientists and crew at sea meditate upon the different kinds of expertise these parties have, on how class hierarchies are often reinforced through divisions between "mental" and "manual" labor. See H. Russell Bernard and Peter D. Killworth, "On the Social Structure of an Ocean-Going

Research Vessel and Other Important Things," *Social Science Research* 2, no. 2 (1973): 145–184.

13. Together, scientists and pilots enact what Edwin Hutchins, in his ethnography of navigation on board a Navy ship, called "cognition in the wild," converting and communicating information through a "cascade of representations" channeled through analog and digital machines and different people's embodied expertise, different sorts of selves, which are here brought into intimate, close-quarters relation. See Edwin Hutchins, *Cognition in the Wild* (Cambridge, Mass.: MIT Press, 1995).

14. This essay only scratches the surface of a more technical history of ROVs (or, in another direction, what Sherry Turkle has called an "inner history," focusing on subjectivity) that could be written. For an instructive primary source, detailing the control system of the Woods Hole Oceanographic Institution's JASON ROV, see David A. Mindell, Dana R. Yoerger, Lee E. Freitag, Louis L. Whitcomb, and Robert L. Eastwood, "JASONTALK: A Standard Remotely Operated Vehicle (ROV) Vehicle Control System," *Proceedings of the IEEE/MTS Oceans Conference, Victoria, B.C., Canada., 1993* (Piscataway, N.J.: IEEE), 253–258. See also Sherry Turkle, ed., *The Inner History of Devices* (Cambridge, Mass.: MIT Press, 2008).

15. Sherry Turkle, *The Second Self: Computers and the Human Spirit* (Cambridge, Mass.: MIT Press. 2005 [1984]), 7–12.

16. Goodwin, "Seeing in Depth," 250.

17. Eva Shawn Hayward, "Jellyfish Optics: Immersion in Marine Techno-Ecology." Paper prepared for the meetings of the Society for Science and Literature, Durham, NC, October 14–17, 2004.

18. The charter:

Cruise Plan, Wednesday, March 26, 2003
Ship: R/V Point Lobos
Chief Scientist: Tamara Robena

Subtract 7 hours during PDT to convert to local time, 8 hours during PST

Scheduled Start Time: Wednesday 26 March 2003 1500 UTC
Scheduled End Time: Thursday 27 March 2003 0030 UTC

Purpose: To visit the 1500 m cold seeps. To collect clams (a new species currently being described) and sediment cores.

Required Equipment: Animal drawer push cores (8 total) clam scoop or gravity sampler Niskin bottles (2)

19. Matters would be different with other ROVs, which do feature force feedback.

20. Media theorists Jay David Bolter and Richard Grusin offer that: "Our culture wants both to multiply its media and to erase all traces of mediation; ideally, it wants to erase its media in the very act of multiplying them." Jay David Bolter and Richard Grusin, *Remediation: Understanding New Media* (Cambridge, Mass.: MIT Press, 1999), 5. As Turkle and her colleagues suggest, "Rather than 'dematerializing' humans or their objects, computer technologies 'remediate' the practices and 'stuff' of science, engineering, and design." Turkle et. al., "Information Technologies and Professional Identity," 39.

21. Marshall McLuhan, *Understanding Media: The Extensions of Man* (New York: McGraw-Hill, 1964), 7, 18.

22. See Turkle, *The Second Self,* and *Life on the Screen: Identity in the Age of the Internet* (New York: Simon and Schuster, 1995).

BUILDINGS AND BIOLOGY

Yanni A. Loukissas, Keepers of the Geometry

1. I have studied architects and their collaborators in numerous professional and academic environments; this essay focuses on the stories of two organizations, addressed here through the pseudonyms Paul Morris Associates and Ralph Jerome Architects. All interviews reported in this essay were conducted between 2002 and 2005. I wish to express my deep gratitude to my informants. I would also like to thank Sherry Turkle for reviewing

and substantially editing multiple versions of this essay and for supervising much of the ethnographic research. Additional thanks to William Porter and Robin Macgregor for reading this essay and providing comments. My work has been partially supported by the National Science Foundation under Grant No. 0220347. Any opinions, findings, and conclusions or recommendations expressed in this material are those of the author and do not necessarily reflect the views of the National Science Foundation. In this essay, the anonymity of all informants and institutions is preserved.

2. CATIA is an acronym for "Computer-Aided Three-Dimensional Interactive Application."

3. Dana Cuff, *Architecture: The Story of Practice* (Cambridge, Mass.: MIT Press, 1991).

4. Magali Sarfatti Larson, *Behind the Postmodern Facade: Architectural Change in Late Twentieth-Century America* (Berkeley: University of California, 1993).

5. Individual architects will, of course, have different personal styles of relating to computation. My analysis here is at the intersection of artifact, personal style, and culturally available roles within the profession. See Sherry Turkle and Seymour Papert, "Epistemological Pluralism and the Revaluation of the Concrete," *Signs* 16, no.1 (Autumn 1990):128–157.

6. In the office of Paul Morris Associates, the responsibility for coordinating all of the geometric modeling work on a given project falls upon one architect, referred to by his peers as the "keeper of the geometry."

7. See Sherry Turkle, *Life on the Screen: Identity in the Age of the Internet* (New York: Simon and Schuster, 1995).

8. In the highly influential office of H. H. Richardson, Richardson himself played the parts of a manager and a model of practice. Richardson was among the first American architects trained at the Ecole des Beaux-Arts in Paris. At the time, there was no such institution in America. Therefore,

Richardson took it upon himself to combine training with practice in his office.

9. See Cuff, *Architecture,* and Donald A. Schön, *The Reflective Practitioner: How Professionals Think in Action* (New York: Basic Books, 1983).

10. Schön, *The Reflective Practitioner,* 80.

11. Reyner Banham, "A Black Box: The Secret Profession of Architecture," in *A Critic Writes: Selected Essays by Reyner Banham* (Berkeley: University of California Press, 1996).

12. Maurice Merleau-Ponty, *Phenomenology of Perception* (New York: Routledge, 2002).

13. Peter Galison, *Image and Logic* (Chicago: University of Chicago Press, 1997), chapter 9.

14. Diana Forsythe, *Studying Those Who Study Us: An Anthropologist in the World of Artificial Intelligence* (Stanford: Stanford University Press, 2001); Lucy A. Suchman, *Plans and Situated Actions: The Problem of Human-Machine Communication* (Cambridge: Cambridge University Press, 1987).

15. The phrase "frozen ideology" was used by William J. Mitchell in a personal conversation I had with him in 2005.

16. Gary Lee Downey, *The Machine in Me: An Anthropologist Sits among Computer Engineers* (New York: Routledge, 1998).

17. Michel Foucault, *Discipline and Punish: The Birth of the Prison* (New York: Vintage Books, 1995), 94.

18. Andrew Abbott, *The System of Professions: An Essay on the Division of Expert Labor* (Chicago: University of Chicago Press, 1988).

Natasha Myers, Performing the Protein Fold
1. Research for this paper was generously supported by an NSF Predoctoral Fellowship (Grant No. 00220347), an NSF Dissertation Improvement Grant

(Award No. 0646267), a Social Sciences and Humanities Research Council of Canada Doctoral Fellowship (Award No. 752-2002-0301). Any opinions, findings, and conclusions or recommendations expressed in this material are those of the author and do not necessarily reflect the views of the National Science Foundation or any other granting agency.

For a more extended discussion of my ideas on models, see Natasha Myers, "Modeling Proteins, Making Scientists: An Ethnography of Pedagogy and Visual Cultures in Contemporary Structural Biology" (PhD dissertation, Massachusetts Institute of Technology, 2007).

2. James K. Griesemer asserts that a history of models must account for how they are performed in practice. See James R. Griesemer, "Three-Dimensional Models in Philosophical Perspective," in *Models: The Third Dimension of Science,* ed. Soraya de Chadarevian and Nick Hopwood (Stanford: Stanford University Press, 2004).

3. See David Kaiser, *Drawing Theories Apart: The Dispersion of Feynman Diagrams in Postwar Physics* (Chicago: University of Chicago Press, 2005); and Maria Trumpler, "Converging Images: Techniques of Intervention and Forms of Representation of Sodium-Channel Proteins in Nerve Cell Membranes," *Journal of the History of Biology* 30 (1997): 55–89.

4. See Kaiser, *Drawing Theories Apart.*

5. Natasha Myers, "Molecular Embodiments and the Body-Work of Modeling in Protein Crystallography," *Social Studies of Science* 38, no. 2 (2008): 163–199, and "Modeling Proteins, Making Scientists."

6. Trumpler, "Converging Images," 55.

7. Ibid., 56.

8. Here I follow Annemarie Mol in her attempt to track practices through their "enactment" rather than through "representations." See Annemarie Mol, *The Body Multiple: Ontology in Medical Practice, Science and Cultural Theory* (Durham, N.C.: Duke University Press, 2002). On models as objects to be "engaged" and "inhabited," see Donna J. Haraway, *Modest_Witness*

@ *Second_Millennium.Femaleman_ Meets_Oncomouse: Feminism and Techno-science* (New York: Routledge, 1997), 135.

9. Myers, "Molecular Embodiments."

10. I am using pseudonyms here.

11. John C. Kendrew, "Myoglobin and the Structure of Proteins," Nobel Lecture, December 11, 1962," in *Nobel Lectures, Chemistry 1942–1962* (Amsterdam: Elsevier, 1964), 676–698.

12. On cybernetic and informatic rhetoric in the history of molecular biology, see Lily E. Kay, *Who Wrote the Book of Life?: A History of the Genetic Code* (Stanford: Stanford University Press, 2000).

13. For a critical reading of the history of the central dogma, see Evelyn Fox Keller, *The Century of the Gene* (Cambridge, Mass.: Harvard University Press, 2000), and Keller, *Refiguring Life: Metaphors of Twentieth-Century Biology* (New York: Columbia University Press, 1995). For an example of how biologists are currently reworking the central dogma, see Enrico S. Coen, *The Art of Genes: How Organisms Make Themselves* (Oxford: Oxford University Press, 1999).

14. Eric Francoeur, "Cyrus Levinthal, the Kluge and the Origins of Interactive Molecular Graphics," *Endeavour* 26, no. 4 (2002): 127–131; and Eric Francoeur and Jerome Segal, "From Model Kits to Interactive Computer Graphics," in *Models,* ed. Soraya de Chadarevian and Nick Hopwood.

15. Cyrus Levinthal, "Molecular Model-Building by Computer," *Scientific American* 214 (1966): 42–52.

16. Van der Waal's radii describe the volume that a particular atom occupies. Van der Waal's forces prevent atoms from occupying the same volumes.

17. See Myers, "Molecular Embodiments."

18. Peter J. Taylor and Ann S. Blum, "Ecosystems as Circuits: Diagrams and the Limits of Physical Analogies," *Biology and Philosophy* 6 (1991):

276. For insights into the role of models in scientific practice see Mary S. Morgan and Margaret Morrison, eds., *Models as Mediators: Perspectives on Natural and Social Science* (Cambridge: Cambridge University Press, 1999).

19. Michael Lynch, "Science in the Age of Mechanical Reproduction: Moral and Epistemic Relations between Diagrams and Photographs," *Biology and Philosophy* 6 (1991): 208.

20. On how abstractions (and so, analogies) can act as "lures" see Isabelle Stengers, "Whitehead and the Laws of Nature," *Salzburger Theologische Zeitschrift (SaThZ)* 2 (1999): 193–207; and Stengers, "A Constructivist Reading of Process and Reality" (unpublished manuscript, n.d.). For Stengers, abstractions, such as analogies and models, are propositions "asking for, and prompting, a 'leap of imagination'; they act as a lure for feeling, for feeling 'something that matters.'" Effective models, like analogies, can produce what Stengers calls an "empirically felt elucidation of our experience."

21. On how models attract "curious hands," see Robert Langridge et al., "Real-Time Color Graphics in Studies of Molecular-Interactions," *Science* 211, no. 4483 (1981): 661–666.

22. Max F. Perutz, "Obituary: Linus Pauling," *Structural Biology* 1, no. 10 (1994): 667–671.

23. See Hans-Jörg Rheinberger, *Toward a History of Epistemic Things: Synthesizing Proteins in the Test Tube* (Stanford: Stanford University Press, 1997).

24. Bruno Latour uses the term "articulation" to examine how researchers' bodies and senses are entrained to laboratory protocols and techniques. Rachel Prentice draws on Latour to theorize a process of "mutual articulation" at work between researchers and their computers in the production of effective computer simulations for teaching surgery and anatomy. See Bruno Latour, "How to Talk about the Body? The Normative Dimensions of Science Studies," *Body and Society* 10, no. 2–3 (2004): 205–229; and Rachel Prentice, "The Anatomy of a Surgical Simulation: The Mutual Articulation

of Bodies in and through the Machine," *Social Studies of Science* 35 (2005): 867–894.

25. Elinor Ochs, Sally Jacobi, and Patrick Gonzales, "Interpretive Journeys: How Physicists Talk and Travel through Graphic Space," *Configurations* 2, no. 1 (1994): 158, 161.

26. On "becoming," see Gilles Deleuze and Félix Guattari, *A Thousand Plateaus: Capitalism and Schizophrenia,* trans. Brian Massumi (Minneapolis: University of Minnesota Press, 1980).

ABOUT THE AUTHORS

William J. Clancey works at the NASA Ames Research Center and Florida Institute of Human & Machine Cognition. He received a BA in Mathematical Sciences from Rice University (1974) and a PhD in Computer Science from Stanford University (1979). Chief Scientist for Human-Centered Computing, he has extensive experience in medical, educational, and financial software and was a founding member of the Institute for Research on Learning. His six books include *Situated Cognition: On Human Knowledge and Computer Representations* and *Contemplating Minds: A Forum for Artificial Intelligence* (coedited with Stephen Smoliar and Mark J. Stefik).

Stefan Helmreich is Associate Professor of Anthropology at MIT. He has written extensively on artificial life, a field dedicated to the computer simulation of living systems, notably in *Silicon Second*

Nature: Culturing Artificial Life in a Digital World. He is the author of a book about how science is reimagining the oceans, *Alien Ocean: Anthropological Voyages in Microbial Seas* (2009).

Yanni A. Loukissas is an architect and researcher who specializes in the social and cultural study of design technologies. He has taught design studio and theory at MIT, Cornell University, and the School of the Museum of Fine Arts in Boston. He received his PhD in Design and Computation at MIT. His dissertation tracks the coevolution of information technologies for simulation and new conceptions of building design. He is also a consultant at Small Design Firm, where he is working on a new art information system for the Metropolitan Museum of Art in New York City. He is currently Visiting Lecturer in the Department of Architecture at Cornell University.

Natasha Myers is an Assistant Professor in the Department of Anthropology and in the Science and Technology Studies Program at York University. As an anthropologist working in the field of science and technology, she examines the lively visual and performance cultures that thrive in contemporary life science laboratories and classrooms.

Sherry Turkle is Abby Rockefeller Mauzé Professor of the Social Studies of Science and Technology at MIT in the Program for Science, Technology, and Society and the founder and director of the MIT Initiative on Technology and Self. Her books include *The Second Self: Computers and the Human Spirit* and *Life on the Screen:*

Identity in the Age of the Internet. She recently edited and wrote introductory essays for a trilogy of books on objects, subjectivity, and ways of knowing, with special emphasis on ways of knowing in science, technology, and design. These are: *Evocative Objects: Things We Think With, Falling for Science: Objects in Mind,* and *The Inner History of Devices.* She is currently writing a book on the new meanings of intimacy in contemporary digital culture.

INDEX